# Cold-Formed Tubular Members and Connections

## Structural Behaviour and Design

# Cold-Formed Tubular Members and Connections

## Structural Behaviour and Design

**Xiao-Ling Zhao**
Monash University, Australia

**Tim Wilkinson**
The University of Sydney, Australia

**Gregory Hancock**
The University of Sydney, Australia

2005

ELSEVIER

AMSTERDAM – BOSTON – HEIDELBERG – LONDON – NEW YORK – OXFORD
PARIS – SAN DIEGO – SAN FRANCISCO – SINGAPORE – SYDNEY – TOKYO

| | | | |
|---|---|---|---|
| ELSEVIER B.V. | ELSEVIER Inc. | **ELSEVIER Ltd** | ELSEVIER Ltd |
| Radarweg 29 | 525 B Street, Suite 1900 | **The Boulevard, Langford Lane** | 84 Theobalds Road |
| P.O. Box 211, 1000 AE | San Diego, | **Kidlington, Oxford** | London |
| Amsterdam, The Netherlands | CA 92101-4495, USA | **OX5 1GB, UK** | WC1X 8RR, UK |




First edition 2005

ISBN: 0-080-4410-17

Printed in Great Britain.

# Preface

Extensive research projects on tubular structures have been carried out in the last 30 years under the direction of CIDECT (International Committee for the Development and Study of Tubular Structures) and IIW (International Institute of Welding) Subcommission XV-E. A series of design guides have been produced by CIDECT to assist practising engineers. Individual steel manufacturers have been involved in numerous research programs on their own products. Professional organisations such as the Australian Steel Institute (formerly the Australian Institute of Steel Construction), The Steel Construction Institute (UK), the American Institute of Steel Construction, the Canadian Institute of Steel Construction and the Architectural Institute of Japan and the Building Centre of Japan have also prepared design aids on designing steel hollow sections. Most of the documents were related mainly to the behaviour and design of hot-rolled tubular sections.

This book describes the structural behaviour and design of cold-formed tubular members and connections. Cold-formed tubes have several special characteristics which differentiate them from hot-rolled tubes such as rounded stress-strain material behaviour, variation of yield stress around the section, larger residual stresses, web crippling of RHS due to the external corner radii that introduce load eccentrically to the webs, interaction of web local buckling and flange local bucking in bending, weld defects in welded thin-walled tubes and their impact on fatigue strength, and challenge for plastic design because of lower ductility. The following topics on cold-formed tubular sections have only received small coverage in the existing design standards, design guides or relevant books: members subjected to bending, compression, combined bending and compression, local buckling under concentrated force, effect of bending on bearing capacity, tension members and welds in thin-walled tubes, welded connections subjected to fatigue loading, effect of concrete-filling and large-deformation cyclic loading on limiting width-to-thickness ratios, fatigue design using hot spot stress method, bolted moment end plate connections and plastic design of portal frames. These topics are addressed in detail in the book.

This book not only summarises the research performed to date on cold-formed tubular members and connections but also provides design examples in accordance with both the Australian Standard AS 4100 and the British Standard BS 5950 Part 1. It is suitable for structural engineers, researchers and university students who are interested in tubular structures.

Chapter 1 deals with the application of cold-formed tubes and the scope of the book. Chapter 2 summarises the manufacturing processes and manufacturing tolerances in various standards. It also presents the material properties of cold-formed tubes including the rounded stress-strain curves, variation of yield stress around the section, residual stresses and fracture toughness. Chapters 3, 4 and 5 are concerned with members subjected to bending, compression and combined bending and compression. The highlights include slenderness limits, flexural-torsional buckling, interaction of local and overall buckling and beam-column behaviour. Chapter 6 discusses RHS members subjected to concentrated forces applied through either a welded brace or a bearing plate. The effect of bending moment on bearing capacity is also presented.

Tension members and welds in thin-walled tubes are covered in Chapter 7. Chapter 8 describes welded connections subjected to fatigue loading. The classification method is discussed in detail. Chapter 9 presents some recent developments in cold-formed tubular members and connections. The highlights are limiting width-to-thickness ratios for concrete-filled tubes and for those subjected to large-deformation cyclic loading, fatigue design using hot spot stress method, bolted moment end plate connections and plastic design of portal frames. Extensive references are given in the book.

We are grateful for the advice on tubular structures from Prof. Jaap Wardenier at Delft University of Technology, Prof. Jeffery Packer at the University of Toronto, Prof. Donald Sherman at The University of Wisconsin-Milwaukee, Prof. Yoshi Kurobane at Kumamoto University and Prof. Paul Grundy at Monash University over the last 10 years.

We appreciated the comments from Dr. Leroy Gardner at Imperial College, London on Chapter 6, Dr. Steve Maddox at TWI, UK and Dr. Alain Nussbaumer at EPFL ICOM, Lausanne on Chapter 8, Dr. Mohamed Elchalakani at Connell Wagner Pty Ltd on Chapters 3 and 9, and Dr. Fidelis Mashiri at Monash University on Chapters 8 and 9. Thanks are given to Dr. Mike Bambach at Monash University for checking the design examples in the book. Prof. David Nethercot at Imperial College, London, Charles King and Abdul Malik at The Steel Construction Institute, UK provided necessary documents regarding BS 5950 Part 1. Prof. Yuji Makino at Kumamoto University provided necessary information regarding JIS standards. We are very grateful to Mr. Robert Alexander at Monash University for preparing most of the diagrams. We wish to thank OneSteel Market Mills for providing the front cover photo and Smorgon Steel Tube Mills for providing the back cover photo. We also wish to thank Keith Lambert, Lorna Canderton and Noel Blatchford at Elsevier Ltd for their advice on the format of the book.

Finally, we wish to thank our families for their support and understanding during the many years that we have been undertaking tubular research, both at The University of Sydney and Monash University, and during the preparation of the book.

Xiao-Ling Zhao, Tim Wilkinson and Gregory Hancock
January 2005

# Table of Contents

# Notation

The following notation is used in this book except for Chapter 8 and Chapter 9 where symbols are defined within the chapters. Where non-dimensional ratios are involved, both the numerator and denominator are expressed in identical units. The dimensional units for length and stress in all expressions or equations are to be taken as millimetres and megapascals ($N/mm^2$) respectively, unless specifically noted otherwise. When more than one meaning are assigned to one symbol, the correct one will be evident from the context in which it is used. Some symbols are not listed here because they are only used in one section and well defined in the local context.

| | |
|---|---|
| $A_e$ | Effective net area |
| $A_{eff}$ | Effective cross-sectional area |
| $A_g$ | Gross cross-sectional area |
| $A_n$ | Net area of a cross-section |
| $B$ | Overall flange width of an RHS |
| $D$ | Outside diameter of a CHS, or<br>Overall depth of an RHS |
| $E$ | Young's modulus of elasticity |
| $F_t$ | Tensile axial force |
| $G$ | Shear modulus of elasticity |
| $I$ | Second moment of area |
| $I_w$ | Warping constant of a cross-section |
| $I_x$ | $I$ about the cross-section major principal $x$-axis |
| $I_y$ | $I$ about the cross-section minor principal $y$-axis |
| $J$ | Torsion constant for a cross-section |
| $L$ | Member length, or<br>Total weld length |
| $L_E$ | Effective length of a member |
| $L_w$ | Weld length defined in Figure 7.1 |
| $M$ | Bending moment, or<br>Specified mass defined in Table 2.3 |
| $M_b$ | Nominal member moment capacity |
| $M_{bx}$ | $M_b$ about major principal $x$-axis |
| $M_c$ | Moment capacity |
| $M_f$ | Capacity of a member subjected to pure bending |
| $M_i$ | Nominal in-plane member moment capacity |
| $M_{ix}$ | $M_i$ about major principal $x$-axis |
| $M_{iy}$ | $M_i$ about minor principal $y$-axis |
| $M_{max}$ | Maximum moment |
| $M_o$ | Elastic flexural-torsional buckling moment |
| $M_{ox}$ | Nominal out-of-plane member moment capacity about major principal $x$-axis |
| $M_p$ | Plastic moment |
| $M_{rx}$ | $M_s$ about major principal $x$-axis reduced by axial force |
| $M_{ry}$ | $M_s$ about minor principal $y$-axis reduced by axial force |
| $M_s$ | Nominal section moment capacity |
| $M_{sx}$ | $M_s$ about major principal $x$-axis |
| $M_{sy}$ | $M_s$ about minor principal $y$-axis |

| | |
|---|---|
| $M_x^*$ | Design bending moment about major principal *x*-axis |
| $M_y$ | Yield moment |
| $M_y^*$ | Design bending moment about minor principal *y*-axis |
| $M_{yz}$ | Yield moment |
| $N_c$ | Nominal member capacity in compression |
| $N_{cy}$ | $N_c$ for member buckling about minor principal *y*-axis |
| $N_s$ | Nominal section capacity of a compression member |
| $N_t$ | Nominal section capacity in tension |
| $N^*$ | Design axial force |
| $P$ | Applied force |
| $P_c$ | Compression resistance |
| $P_f$ | Capacity of a member subjected to concentrated force only |
| $P_L$ | Longitudinal shear capacity per unit length of weld |
| $P_T$ | Transverse shear capacity per unit length of weld |
| $P_{web\ buckling}$ | Web buckling capacity |
| $R$ | Rotation capacity |
| $R_b$ | Nominal bearing capacity of a web |
| $R_{bb}$ | Nominal bearing buckling capacity |
| $R_{by}$ | Nominal bearing yield capacity |
| $R_{req}$ | Required rotation capacity |
| $S$ | Plastic section modulus |
| $S_{eff}$ | Effective plastic section modulus |
| $S_{rx}$ | Reduced plastic section modulus about the major axis |
| $S_{ry}$ | Reduced plastic section modulus about the minor axis |
| $S_x$ | Plastic section modulus about the major axis |
| $S_y$ | Plastic section modulus about the minor axis |
| $S_o$ | Cross-sectional area of a tensile coupon |
| $S^*$ | Design action effect |
| $T$ | Gusset plate thickness |
| $T_r$ | Shear lag resistance |
| $V$ | Twist defined in Figure 2.2 |
| $V_r$ | Parent metal shear resistance |
| $Z$ | Elastic section modulus |
| $Z_c$ | $Z_e$ for a compact section |
| $Z_e$ | Effective section modulus |
| $Z_{eff}$ | Effective section modulus |
| $Z_x$ | Elastic section modulus about the major axis |
| $Z_y$ | Elastic section modulus about the minor axis |
| | |
| $a$ | Weld throat thickness |
| $b$ | Overall flange width of an RHS |
| $b_e$ | Effective width |
| $b_0$ | Chord flange width |
| $b_1$ | Brace flange width |
| $c_m$ | Factor for unequal moments |
| $d$ | Overall depth of an RHS |
| $d_o$ | Outside diameter of a CHS |
| $e$ | Eccentricity defined in Figure 6.13 |
| $f_o$ | Elastic local buckling stress |
| $f_u$ | Ultimate tensile strength |

| | |
|---|---|
| $f_{uw}$ | Weld metal strength |
| $f_y$ | Tensile yield stress |
| $h_0$ | Chord web depth |
| $h_1$ | Brace web depth |
| $k$ | Plate buckling coefficient |
| $k_e$ | Member effective length factor |
| $k_f$ | Form factor |
| $k_l$ | Load height factor |
| $k_r$ | Lateral rotation restraint factor |
| $k_t$ | Twist restraint factor |
| $l$ | Member length |
| $l_e$ | Effective length of a member |
| $n$ | Number of tensile bolts |
| $p_b$ | Bending strength |
| $p_c$ | Compressive strength |
| $p_{cs}$ | Compressive strength for class 4 slender cross-section |
| $p_w$ | Design strength of a fillet weld |
| $p_y$ | Yield stress |
| $r$ | Internal corner radius of an RHS |
| $r_{ext}$ | External corner radius of an RHS |
| $r_x$ | Radius of gyration about major principal axis |
| $r_y$ | Radius of gyration about minor principal axis |
| $s$ | Leg length of a fillet weld |
| $t$ | Tube wall thickness |
| $t_0$ | Chord wall thickness |
| $t_1$ | Brace wall thickness |
| $w$ | Distance between welds measured around the perimeter of the tube defined in Section 7.4.1 |
| | |
| $\alpha$ | Shear lag reduction factor |
| $\alpha_b$ | Compression member section constant |
| $\alpha_c$ | Compression member slenderness reduction factor, or Reduction factor defined in Section 6.2.2 |
| $\alpha_m$ | Moment modification factor |
| $\alpha_p$ | Coefficient used to calculate the nominal bearing yield capacity ($R_{by}$) |
| $\alpha_s$ | Slenderness reduction factor |
| $\beta$ | Ratio of the brace width to the chord width ($b_1/b_0$) for RHS |
| $\beta_m$ | Ratio of smaller to larger bending moment at the ends of a member, or Ratio of end moment to fixed end moment |
| $\varepsilon$ | Constant $(250/p_y)^{0.5}$ in AS 4100, or Constant $(275/p_y)^{0.5}$ in BS 5950 Part 1 |
| $\varepsilon'$ | Constant $(235/p_y)^{0.5}$ in Eurocode 3 |
| $\varepsilon_u$ | Strain at the ultimate tensile strength |
| $\varepsilon_y$ | Yield strain |
| $\phi$ | Capacity factor |
| $\gamma_f$ | Partial factor for loads |
| $\gamma_m$ | Partial factor for material strength |
| $\kappa$ | Curvature |
| $\kappa_1$ | Curvature when moment drops below $M_p$ defined in Figure 3.5 |

| | |
|---|---|
| $\kappa_p$ | Plastic curvature |
| $\lambda$ | Slenderness ratio |
| $\lambda_e$ | Plate element slenderness |
| $\lambda_{ep}$ | Plate element plasticity slenderness limit |
| $\lambda_{ey}$ | Plate element yield slenderness limit |
| $\lambda_n$ | Modified compression member slenderness |
| $\lambda_s$ | Section slenderness |
| $\lambda_{sp}$ | Section plasticity slenderness limit |
| $\lambda_{sy}$ | Section yield slenderness limit |
| $\lambda_w$ | $\lambda_e$ for the web in compression only |
| $\lambda_{wy}$ | $\lambda_{ey}$ for the web in compression only |
| $\nu$ | Poisson's ratio |

| | |
|---|---|
| AIJ | Architectural Institute of Japan |
| AISC | American Institute of Steel Construction |
| ASI | Australian Steel Institute |
| AWS | American Welding Society |
| BSI | British Standard Institution |
| CHS | Circular Hollow Section |
| CIDECT | International Committee for the Development and Study of Tubular Structures |
| DEn | Department of Energy |
| EC3 | Eurocode 3 |
| IIW | International Institute of Welding |
| kN | Kilonewton |
| m | Metre |
| mm | Milimetre |
| MF | Magnification Factor |
| MPa | Megapascal (N/mm$^2$) |
| RHS | Rectangular Hollow Section |
| SCF | Stress Concentration Factor |
| SCI | The Steel Construction Institute, UK |
| SHS | Square Hollow Section |

# Chapter 1: Introduction

## 1.1 Application of Cold-Formed Tubular Sections

Cold-formed structural members are being used more widely in routine structural design as the world steel industry moves from the production of hot-rolled section and plate to coil and strip, often with galvanised and/or painted coatings. Steel in this form is more easily delivered from the steel mill to the manufacturing plant where it is usually cold-rolled into open and closed section members. Structural steel hollow sections (commonly called tubular sections) may be manufactured to a large variety of material design specifications and standards in different parts of the world. They may also be manufactured by a variety of manufacturing processes. The usual process of manufacture is to form hot or cold-rolled steel strip to a circular shape then to join the abutting edges by an electric resistance weld (ERW) or submerged arc weld before final forming to the design shapes, usually circular hollow sections (CHS), square hollow sections (SHS) or rectangular hollow sections (RHS). In some applications of tubular members, the sections are in-line galvanised with a subsequent enhancement of the tensile properties.

Cold-formed structural steel hollow sections are now permitted to all the major structural steel design standards in the world including the American Institute of Steel Construction LRFD specification (AISC 1999), British Standard BS5950 Part 1 (BSI 2000), Australian Standard AS 4100 (Standards Australia 1998), Canadian Standard CSA-S16-01 (2001) and the proposed Eurocode 3 (EC3 2003). There is a potential increased market in South East Asia, such as in mainland China, Hong Kong and Singapore, for cold-formed tubular sections.

In Australia, of the approximately one million tonnes of structural steel used each year, 125,000 tonnes is used for cold-formed open sections such as purlins and girts and 400,000 tonnes is used for tubular members. In Australia, the total quantity of cold-formed products now exceeds the total quantity of hot-rolled products. About one million tonnes of cold-formed square hollow sections were produced in Japan in 2002.

Cold-formed tubular sections are widely used as structural members in steel construction (columns, beams, truss members, scaffoldings), in the transportation industry (bus frames, long distance car carriers), for agricultural equipment (ploughs, transporters), for highway equipment (hand rails, guardrails, pedestrian bridges), for mechanical members (construction machinery, machinery frames) and for recreational structures. A few examples are shown in Figure 1.1.

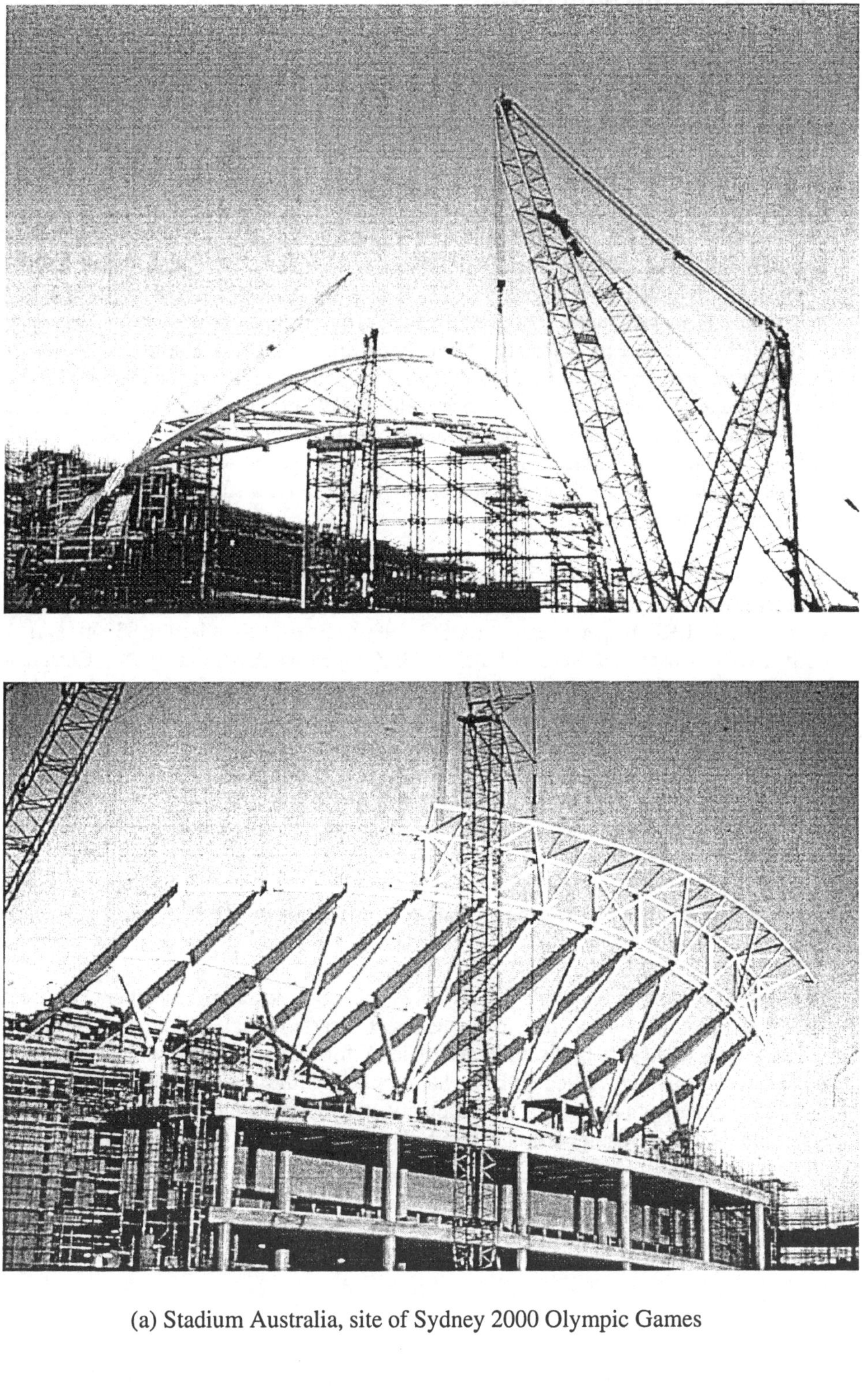

(a) Stadium Australia, site of Sydney 2000 Olympic Games

(b) Federation Square in Melbourne

(c) Kansai International Airport in Osaka

(d) Floor System (photo courtesy of OneSteel Market Mills)

(e) Roof System (photo courtesy of Smorgon Steel Tube Mills)

(f) Scaffolding (photo courtesy of OneSteel Market Mills)

(g) Bus Frame (photo courtesy of OneSteel Market Mills)

(h) Pedestrian Bridge (photo courtesy of OneSteel Market Mills)

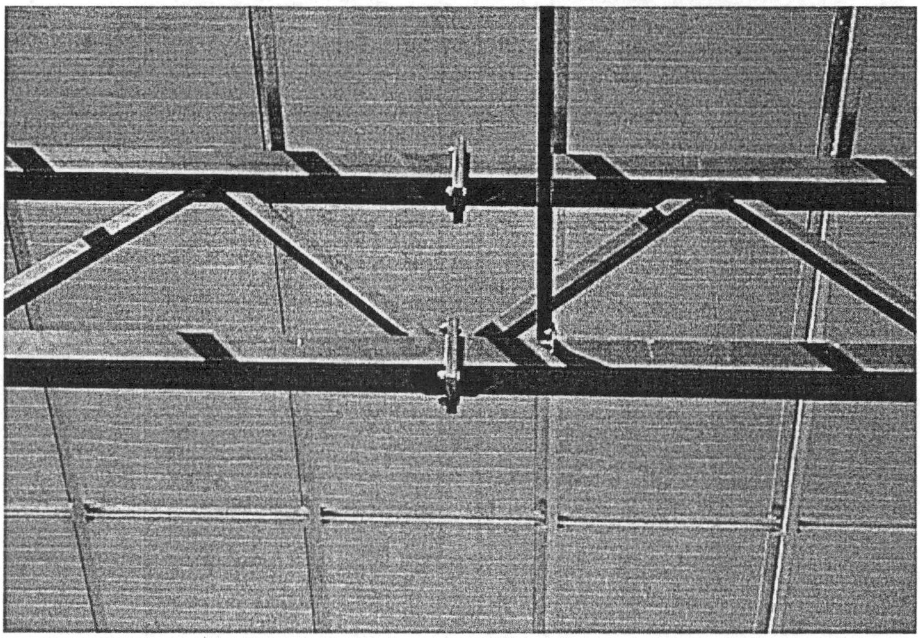

(i) Truss (photo courtesy of Smorgon Steel Tube Mills)

(j) Space Frame (photo courtesy of Smorgon Steel Tube Mills)

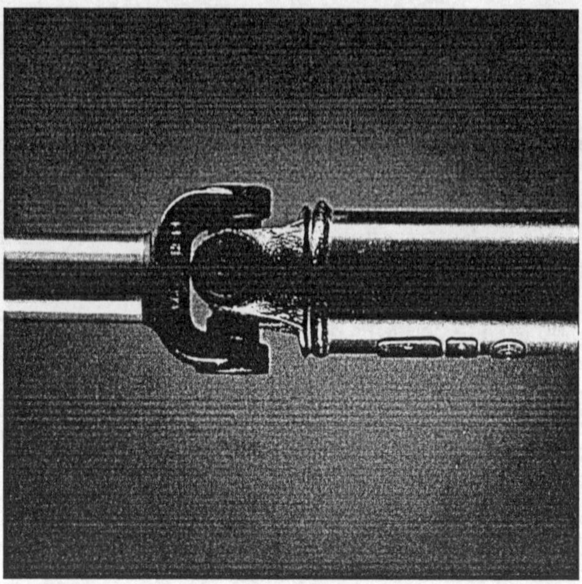

(k) Drive Shaft (photo courtesy of OneSteel Market Mills)

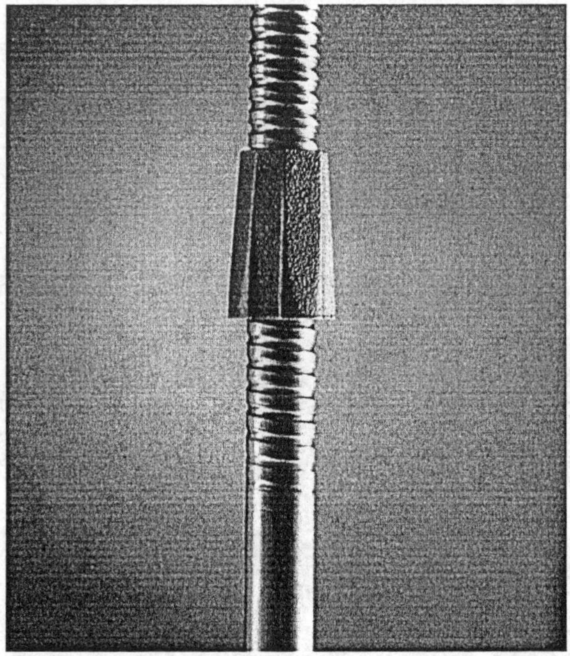

(l) Rock Bolt (photo courtesy of OneSteel Market Mills)

Figure 1.1 Applications of cold-formed tubular sections

## 1.2 International Standards

### *1.2.1 Manufacturing Standards for Cold-Formed Tubular Sections*

In Australia, structural steel hollow sections are normally produced to the Australian Standard AS 1163 (Standards Australia 1991). They are all cold-formed and usually have stress grades of 250 MPa (called C250), 350 MPa (called C350) and 450 MPa (called C450). The most common grade is C350 which has the yield strength enhanced from 300 MPa to 350 MPa during the forming process. The C450 grade is often achieved by in-line galvanizing but may be achieved by alloying elements in the steel feed. Cold-formed structural steel hollow sections are produced to EN 10219 (ENV 1992) in Europe, ASTM A500 (ASTM 1993) in the USA and G3444/G3466 (JIS 1988a, 1988b) in Japan respectively. Detailed comparisons are presented in Section 2.1.

### *1.2.2 Design Standards for Cold-Formed Steel Structures*

Light gauge cold-formed steel structures are designed to AS/NZ S4600 (Standards Australia 1996) in Australia, NAS (2002) in the USA, CSA-S136-01 (2001) in Canada, BS5950 Part 5 (BSI 1998) in the UK and Eurocode 3 Part 1.3 (EC3 2004) in Europe. These standards are mainly developed for cold-formed open sections and sheeting with thickness less than 3 mm (1/8 inch) or 4.6 mm (3/16 inch) although they can be applied up to 25 mm (1 inch) in some cases.

### *1.2.3 Design Standards for Steel Structures that Include Cold-Formed Tubular Sections*

The Australian Standard for the design of steel structures AS 4100 was first published in limit states format in 1990. It was developed mainly for hot-rolled members but permitted the use of cold-formed tubular members to AS 1163. Previously, cold-formed tubular members had been permitted to be designed to the permissible stress steel structures design standard AS 1250 (Standards Australia 1981)) since an amendment in 1982. However, research on cold-formed tubular members was limited in many areas, and so a significant research program was undertaken in the last 25 years. Most of the research outcomes have now been incorporated in Australian Standard AS 4100-1998 and the New Zealand Standard NZS 3404 (1997). The British Standard BS 5950 Part 1 included cold-formed tubular members for the first time in 2000. The American Institute of Steel Construction permissible stress and limit state specifications have allowed cold-formed tubular members since the 1969 edition and 1986 edition, respectively. These AISC Specifications are currently being merged into one document (Lindsey 2003).

Cold-formed tubular sections can now be designed to the mainstream steel structures standards, e.g. AS 4100 in Australia, NZS 3404 in New Zealand, BS 5950 Part 1 in the UK, AISC LRFD-1999 in the USA, CSA-S16-01 in Canada, Eurocode 3 Part 1.1 (2003) in Europe and AIJ (1990a) in Japan, although these standards are mainly applied to traditional sections such as I-sections with thickness larger than 3 mm (1/8 inch) or 4.6 mm (3/16 inch).

### *1.2.4 Recent Design Manuals/Books Published by Professional Organisations*

Extensive research projects on tubular structures were carried out in the last 30 years under the direction of CIDECT (International Committee for the Development and Study of Tubular Structures) and IIW (International Institute of Welding) Subcommission XV-E. Ten international symposia on tubular structures have been held since 1984 (IIW 1984, Kurobane and Makino 1986, Niemi and Mäkeläinen 1989, Wardenier and Panjeh Shahi 1991, Coutie and Davies 1993, Grundy, Holgate and Wong 1994, Farkas and Jámai 1996, Choo and van der Vegte 1998, Puthli and Herion 2001, Jaurrieta et al 2003). A series of design guides have been produced by CIDECT to assist practising engineers. They are:

1 CIDECT Design Guide No.1: Design Guide for Circular Hollow Section (CHS) Joints under Predominantly Static Loading (Wardenier et al 1991)
2 CIDECT Design Guide No.2: Structural Stability of Hollow Sections (Rondal et al 1996)
3 CIDECT Design Guide No.3: Design Guide for Rectangular Hollow Section (RHS) Joints under Predominantly Static Loading (Packer et al 1996)
4 CIDECT Design Guide No. 4: Design Guide for Structural Hollow Section Columns Exposed to Fire (Twilt et al 1996)
5 CIDECT Design Guide No. 5: Design Guide for Concrete Filled Hollow Section Columns under Static and Seismic Loading (Bergmann et al 1995)
6 CIDECT Design Guide No. 6: Design Guide for Structural Hollow Sections in Mechanical Applications (Wardenier et al 1995)

7 CIDECT Design Guide No. 7: Design Guide for Fabrication, Assembly and Erection of Hollow Section Structures (Dutta et al 1998)
8 CIDECT Design Guide No. 8: Design Guide for Circular and Rectangular Hollow Section Welded Joints under Fatigue Loading (Zhao et al 2001)
9 CIDECT Design Guide No. 9: Design Guide for Structural Hollow Section Column Connections (Kurobane et al 2005).

A brief summary on Design Guides 1 to 7 was given by Packer (2000). The Design Guide No.8 focuses on the hot spot stress method, which takes into account most of the influencing factors on fatigue particularly at complex 2D and 3D welded connections. It uses various parametric formulae to calculate the so-called "hot spot stress", which in turn is used to determine the fatigue life of the joint under investigation. The Design Guide No.9 contains design details for beam-to-column connections and end-to-end connections.

Professional organisations such as the Australian Steel Institute (formerly the Australian Institute of Steel Construction), The Steel Construction Institute (UK), the American Institute of Steel Construction, the Canadian Institute of Steel Construction and the Architectural Institute of Japan and the Building Centre of Japan have also prepared design aids on designing steel hollow sections. Some of the documents are listed here:

- Pre-engineered Connections for Structural Steel Hollow Sections (ASI 1997).
- Design Capacity Tables for Structural Steel – Volume 2: Hollow Sections (ASI 1999).
- Steelwork Design Guide to BS 5950-1:2000, Volume 1, Section Properties and Member Capacities, The Steel Construction Institute, UK (SCI 2002)
- Load and Resistance Factor Design Specification for Steel Hollow Structural Sections (AISC 2000).
- Standard for Limit State Design of Steel Structures, Architectural Institute of Japan, Tokyo (AIJ 1990a).
- Recommendations for the Design and Fabrication of Tubular Structures in Steel, Architectural Institute of Japan, Tokyo (AIJ 1990b).
- Design and Fabrication Manual for Cold-Formed Square Tubes, The Building Centre of Japan, Tokyo (BCJ 1996).
- Hollow Structural Sections Connection Manual, American Institute of Steel Construction (AISC 1997)

### *1.2.5 Other Related Books*

The following books are related to tubular structures or cold-formed steel structures:

- Hollow Section Joints by Wardenier (1982) summarised the research work on tubular structures before 1982.
- Design of Welded Tubular Connections – Basis and Use of AWS Code Provisions by Marshall (1992) summarised the design procedures for welded tubular connections in accordance with AWS code.

- Design of Cold-Formed Steel Structures by Hancock (1998) presented design procedures for cold-formed steel structures in accordance with AS/NZS 4600 (1996).
- Mechanics of Concrete Filled Steel Tubes by Han and Zhong (1996) studied the mechanics of concrete filled steel tubes under static loading.
- Hollow Structural Section Connections and Trusses by Packer and Henderson (1997) presented the up-to-date design rules on tubular connections and trusses.
- Tubular Structures in Architecture by Eekhout (1996) described the design possibilities of tubular structures in Architecture applications.
- Hollow Sections in Structural Applications by Wardenier (2001) served as an introduction book for students in Structural and Civil Engineering.
- Cold-Formed Steel Structures to the AISI Specification by Hancock, Murray and Ellifritt (2001) presented design procedures for cold-formed steel structures in accordance with the AISI 1996 specification.
- Structures with Hollow Sections by Dutta (2002) summarized most of the work in CIDECT Design Guides 1 to 7.

## 1.3 Layout of the Book

The following topics on cold-formed tubular sections have only received small coverage in the above mentioned design standards, design guides or relevant books: members subjected to bending, compression, combined bending and compression, local buckling under concentrated force, effect of bending on bearing capacity, tension members and welds in thin-walled tubes, welded connections subjected to fatigue loading, effect of concrete-filling and large-deformation cyclic loading on limiting width-to-thickness ratios, fatigue design using hot spot stress method, bolted moment end plate connections and plastic design of portal frames. These topics are addressed in detail in this book.

This book not only summarises the research performed to date on cold-formed tubular members and connections but also provides design examples in accordance with both the Australian Standard AS 4100 and the British Standard BS 5950 Part 1.

Chapters 1 and 2 outline the application, manufacturing and special characteristics of cold-formed tubular sections. Cold-formed tubular members are covered in Chapter 3 (bending), Chapter 4 (compression), Chapter 5 (combined compression and bending) and Chapter 6 (subject to concentrated forces). Tension members and welds in thin-walled tubes are covered in Chapter 7. Chapter 8 describes welded connections subjected to fatigue loading. Chapter 9 presents the recent developments including limiting width-to-thickness ratios for concrete-filled tubes and for those subjected to large-deformation cyclic loading, fatigue design using the hot spot stress method, bolted moment end plate connections and plastic design of portal frames.

## 1.4 References


1. AIJ (1990a), Standard for Limit State Design of Steel Structures, Architectural Institute of Japan, Tokyo, Japan
2. AIJ (1990b), Recommendations for the Design and Fabrication of Tubular Structures in Steel, Architectural Institute of Japan, Tokyo, Japan
3. AISC (1997), Hollow Structural Sections Connections Manual, American Institute of Steel Construction, Chicago, Illinois, USA
4. AISC (1999), Load and Resistance Factor Design Specification for Structural Steel Buildings, American Institute of Steel Construction, Chicago, Illinois, USA
5. AISC (2000), Load and Resistance Factor Design Specification for Steel Hollow Structural Sections, American Institute of Steel Construction, Chicago, Illinois, USA
6. ASI (1997), Pre-engineered Connections for Structural Steel Hollow Sections, Australian Steel Institute, Sydney, Australia
7. ASI (1999), Design Capacity Tables for Structural Steel – Volume 2: Hollow Sections, Australian Steel Institute, Sydney, Australia
8. ASTM (1993), Standards Specification for Cold-Formed Welded and Seamless Carbon Steel Structural Tubing in Rounds and Shapes, American Society for Testing Materials ASTM A500, USA
9. BCJ (1996), Design and Fabrication Manual for Cold-Formed Square Tubes, The Building Centre of Japan, Tokyo, Japan
10. Bergmann, R., Matsui, C., Meinsma, C. and Dutta, D. (1995), *Design Guide for Concrete Filled Hollow Section Columns under Static and Seismic Loading*, TÜV-Verlag, Köln, Germany
11. BSI (2000), Structural use of Steelwork in Building, BS 5950, Part 1, British Standard Institution, London, UK
12. BSI (1998), Structural use of Steelwork in Building, BS 5950, Part 5, British Standard Institution, London, UK
13. Choo, S. and van der Vegte, G.J. (1998), Tubular Structures VIII, *Proceedings*, 8th International Symposium on Tubular Structures, Singapore, Balkema, Rotterdam, The Netherlands
14. CSA-S16 (2001), Steel Structures for Buildings (Limit State Design), CSA-S16-01, Canadian Standards Association, Toronto, Ontario, Canada
15. CSA-S136 (2001), Cold-Formed Steel Structural Members, CSA-S136-01, Canadian Standards Association, Toronto, Ontario, Canada
16. Coutie, M.G. and Davies, G. (1993), Tubular Structures V, *Proceedings*, 5th International Symposium on Tubular Structures, Nottingham, UK, E & FN Spon, London, UK
17. Dutta, D., Wardenier, J., Yeomans, N., Sakae, K., Bucak, Ö and Packer, J.A. (1998), *Design Guide for Fabrication, Assembly and Erection of Hollow Section Structures*, TÜV-Verlag, Köln, Germany
18. Dutta, D. (2000), *Structures with Hollow Sections*, Ernst & Eohn, Darmstadt, Germany
19. Eekhout, M. (1996), *Tubular Structures in Architecture*, Delft University of Technology, Delft, The Netherlands
20. EC3 (2003), Eurocode 3: Design of Steel Structures – Part 1.1: General Rules and Rules for Buildings, prEN 1993-1-1:2003, November 2003, European Committee for Standardization, Brussels, Belgium

21. EC3 (2004), Eurocode 3: Design of Steel Structures – Part 1.3: Supplementary Rules for Cold-Formed Members and Sheeting, EN 1993-1-3: 2004, 1 March 2004, European Committee for Standardization, Brussels, Belgium
22. ENV (1992), European Committee for Standardization, European Pre-Standard ENV 10219, Cold-Formed Welded Structural Hollow Sections of Non-Alloyed and Fine Grained Steels, Part 1 Technical Delivery condition, Part 2 Tolerances, Dimensions and Section Properties, British Standards Institution, London, UK
23. Farkas, J. and Jámai, K. (1996), Tubular Structures VII, *Proceedings*, 7th International Symposium on Tubular Structures, Miskolc, Hungary, Belkema, Rotterdam, The Netherlands
24. Grundy, P., Holgate, A. and Wong, B. (1994), Tubular Structures VI, *Proceedings*, 6th International Symposium on Tubular Structures, Melbourne, Balkema, Rotterdam, The Netherlands
25. Han, L.H. and Zhong, S.T. (1996), *Mechanics of Concrete Filled Steel Tubes*, Dalian University of Technology Press, Dalian, P.R. China (in Chinese)
26. Hancock, G.J. (1998), *Design of Cold-Formed Steel Structures*, 3rd edition, Australian Institute of Steel Construction, Sydney, Australia
27. Hancock, G.J., Murray, T and Ellifritt, D. (2001), *Cold-Formed Steel Structures to the AISI Specification*, Marcel Dekker, Inc., New York, USA
28. IIW (1984), Welding of Tubular Structures, *Proceedings*, 1st International Symposium on Tubular Structures, Boston, Pergamon Press, Oxford, UK
29. Jaurrieta, M.A., Alonso, A. and Chica, J.A. (2003), Tubular Structures X, *Proceedings*, 10th International Symposium on Tubular Structures, Madrid, Spain, Balkema, Lisse, The Netherlands
30. JIS (1988a), Carbon Steel Tubes for General Structural Purposes, Japanese Industrial Standard, G3444, Tokyo, Japan
31. JIS (1988b), Carbon Steel Square Pipe for General Structural Purposes, Japanese Industrial Standard, G3466, Tokyo, Japan
32. Kurobane, Y. and Makino, Y. (1986), Safety Criteria in Design of Tubular Structures, *Proceedings*, 2nd International Symposium on Tubular Structures, Tokyo, Japan, Architectural Institute of Japan, Tokyo, Japan
33. Kurobane, Y., Packer, J.A., Wardenier, J. and Yeomans, N. (2005), *Design Guide for Structural Hollow Section Column Connections*, TÜV-Verlag, Köln, Germany
34. Lindsey, S.D. (2003), Future Directions of AISC Specifications for Steel Buildings, *Practice Periodical on Structural Design and Construction*, ASCE, 8(3), pp 130-132
35. Marshall, P.W. (1992), *Design of Welded Tubular Connections – Basis and Use of AWS Code Provisions*, Elsevier Science Publishers, Amsterdam, The Netherlands
36. NAS (2002), North American Specification for the Design of Cold-Formed Steel Structural Members, American Iron and Steel Institute, Washington D.C, USA
37. Niemi, E. and Mäkeläinen, P. (1989), Tubular Structures III, *Proceedings*, 3rd International Symposium on Tubular Structures, Lappeenranta, Finland, Elsevier Applied Science, London, UK
38. NZS (1997), Steel Structures Standard, NZS 3404, Part 1, Standards New Zealand, Wellington, New Zealand
39. Packer, J.A., Wardenier, J., Kurobane, Y., Dutta, D. and Yeomans, N. (1992), *Design Guide for Rectangular Hollow Section (RHS) Joints under Predominantly Static Loading*, TÜV-Verlag, Köln, Germany

40. Packer, J.A. and Henderson, J.E. (1997), *Hollow Structural Section Connections and Trusses*, Canadian Institute of Steel Construction, Ontario, Canada
41. Packer, J.A. (2000), Tubular Construction, *Progress in Structural Engineering and Materials,* 2(1), pp 41-49
42. Puthli, R.S. and Herion, S. (2001), Tubular Structures IX, *Proceedings*, 9$^{th}$ International Symposium on Tubular Structures, Dusseldorf, Germany, Balkema, Lisse, The Netherlands
43. Rondal, J., Wurker, K.G., Dutta, D., Wardenier, J. and Yeomans, N. (1992), *Structural Stability of Hollow Sections*, TÜV-Verlag, Köln, Germany
44. SCI (2002), Steelwork Design Guide to BS 5950-1:2000, Volume 1, Section Properties and Member Capacities, The Steel Construction Institute, UK
45. Standards Australia (1981), Steel Structures Code, Australian Standard AS 1250, Standards Australia, Sydney, Australia
46. Standards Australia (1991), Structural Steel Hollow Sections, Australian Standard AS 1163, Standards Australia, Sydney, Australia
47. Standards Australia (1996), Cold-Formed Steel Structures, Australian Standard AS/NZS 4600, Standards Australia, Sydney, Australia
48. Standards Australia (1998), Steel Structures, Australian Standard AS 4100, Standards Australia, Sydney, Australia
49. Twilt, L., Hass, R., Klingsch, W., Edwards, M. and Dutta, D. (1996), *Design Guide for Structural Hollow Section Columns Exposed to Fire*, TÜV-Verlag, Köln, Germany
50. Wardenier, J. (1982), *Hollow Section Joints*, Delft University Press, Delft, The Netherlands
51. Wardenier, J. and Panjeh Shahi, E. (1991), Tubular Structures IV, *Proceedings*, 4$^{th}$ International Symposium on Tubular Structures, Delft, The Netherlands, Delft University Press, The Netherlands
52. Wardenier, J. (2001), *Hollow Sections in Structural Applications*, CIDECT, The Netherlands
53. Wardenier, J., Kurobane, Y., Packer, J.A., Dutta, D. and Yeomans, N. (1991), *Design Guide for Circular Hollow Section (CHS) Joints under Predominantly Static Loading*, TÜV-Verlag, Köln, Germany
54. Wardenier, J., Dutta, D., Yeomans, N., Packer, J. A. and Bucak, Ö. (1995), *Design Guide for Structural Hollow Sections in Mechanical Applications*, TÜV-Verlag, Köln, Germany
55. Zhao, X.L., Herion, S., Packer, J.A., Puthli, R., Sedlacek, G., Wardenier, J., Weynand, K., van Wingerde, A. and Yeomans, N. (2001), *Design Guide for Circular and Rectangular Hollow Section Welded Joints under Fatigue Loading*, TÜV-Verlag, Köln, Germany

# Chapter 2: Cold-Formed Tubular Sections

## 2.1 Manufacturing Processes

Cold-formed tubular sections are manufactured in accordance with different standards, e.g. the Australian Standard AS 1163 (Standards Australia 1991) in Australia, EN 10219 (ENV 1992) in Europe, ASTM A500 (ASTM 1993) in the USA and G3444/G3466 (JIS 1988a, 1988b) in Japan. A comparison is given in Table 2.1 with the methods of manufacture specified in the different standards and specifications.

Table 2.1 Method of manufacture

| Standard | Method of Manufacture |
|---|---|
| AS 1163 | Hollow sections formed and shaped at ambient temperature from a single strip of steel, both edges of which are continuously welded by either the electric resistance or submerged arc process. |
| EN 10219 | Cold-formed without subsequent heat treatment. Shall be manufactured by electric resistance welding or submerged arc welding without subsequent heat treatment. |
| ASTM A500 | Welded tubing made from flat-rolled steel by the electric resistance welding process. May be stress relieved or annealed as is considered necessary by the manufacturer to conform to the specification. Grade D tubing shall be heat treated at 610°C. |
| G3444/ G3466 | Shall be manufactured by seamless process, electric resistance welding, butt welding or arc welding. They shall usually be furnished as-manufactured without heat treatment. |

A typical manufacturing process is shown in Figure 2.1. Brief explanations of each step in Figure 2.1 are given as follows:

Step 1: Uncoiling and Joining Coils
The coils are prepared for the start of the manufacturing process by uncoiling and levelling. The edges are trimmed and the flat steel is then slit into the required widths to suit the final section sizes. The ends of the coils are joined transversely by welding. A looper tunnel is normally used to allow a loop of steel strip to feed the mill while the coil joining operation takes place.

Step 2: Forming
A series of rollers form the steel strip into a circular shape. The strip is not artificially heated during the gradual cold-forming process.

Step 3: Welding
When the edges of the formed circular shape are pushed together by squeeze rollers, they are welded using ERW (Electric Resistance Welding) to form a circular hollow section (CHS). The external weld bead is removed by a weld trimmer.

Step 4: Sizing and Shaping
A series of rollers (called stages) are used to turn the CHS tube into a square or rectangular hollow section (SHS or RHS) or to size the CHS accurately.

Step 5: Cutting and Bundling
The finished tubular sections are cut to specified lengths using an electrically controlled cut-off machine. They are then packed and despatched.

It should be noted that sometimes in-line painting and in-line galvanising are steps in the manufacturing process. The painting offers protection for steel tubes during transport, handling and fabrication. The in-line galvanising not only increases the corrosion resistance but may also enhance the strength of the steel tubes. The painting step is between the shaping and cutting operations. The in-line galvanising step occurs before sizing and shaping.

## 2.2 Manufacturing Tolerances

### *2.2.1 Tolerance Values*

The manufacturing tolerances specified in various manufacturing standards (AS 1163, EN 10219, ASTM A500 and G3444/G3466) are compared in Tables 2.2 and 2.3. The tolerances in cross-section (outside dimension, thickness, external corner radii) are presented in Table 2.2 while those in length, straightness, twist and mass are listed in Table 2.3.

Basic dimensions of cold-formed tubes are defined in Figure 2.2. The symbols defined in Figure 2.2 are used throughout this book. The twist ($V$ in Table 2.3) is defined in Figure 2.3. The term $M$ in the last row of Table 2.3 stands for the specified mass. SHS (square hollow section) is a special case of RHS (rectangular hollow section) when $b$ equals $d$.

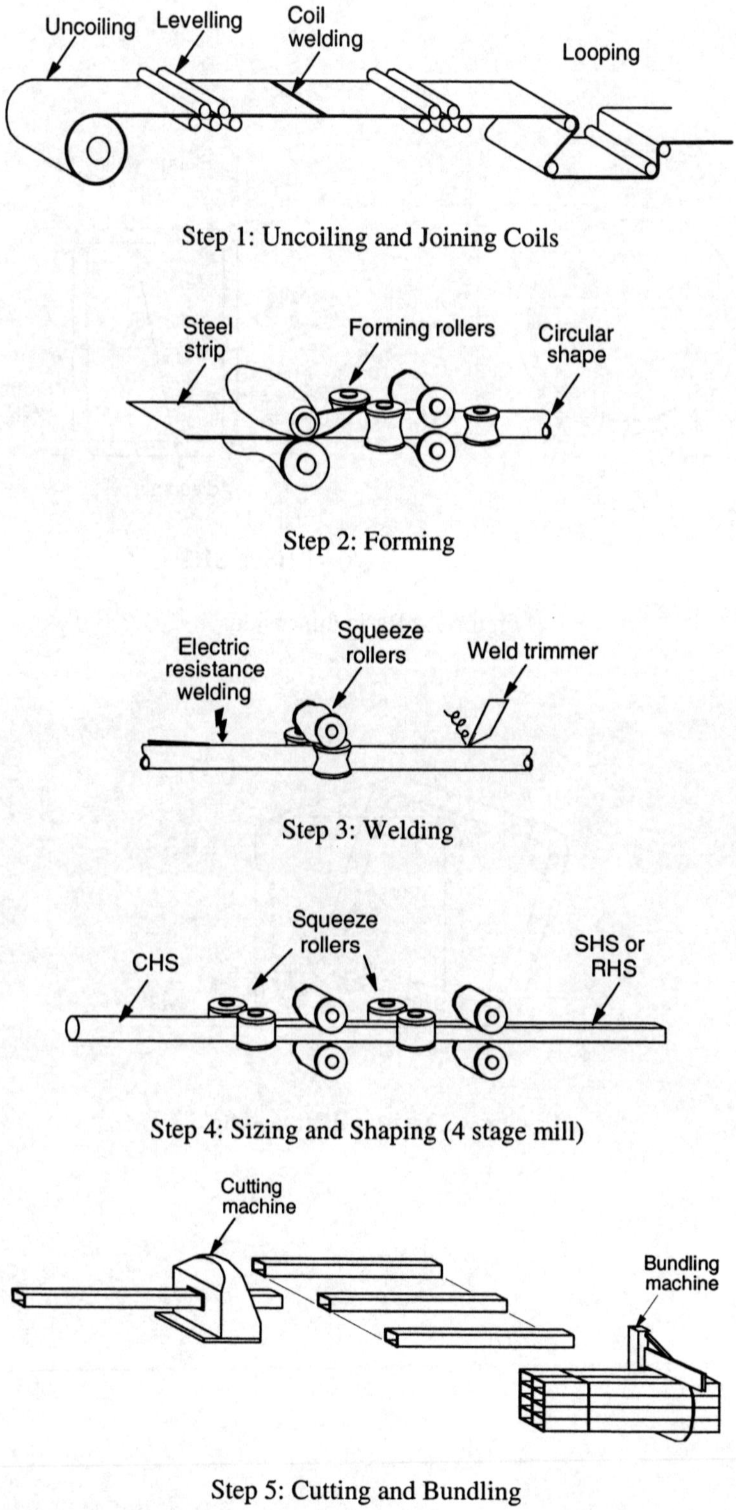


Figure 2.1 Schematic illustration of major steps in typical manufacturing process

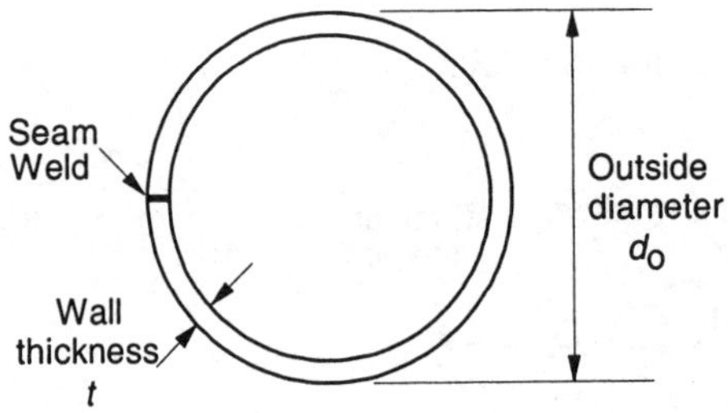


(a) CHS

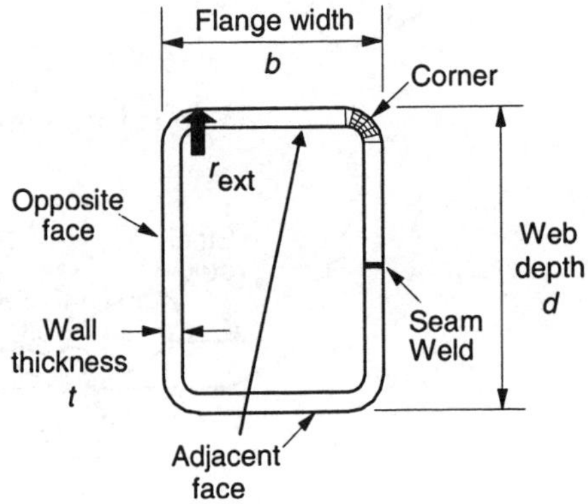


(b) RHS or SHS

Figure 2.2 Basic dimensions

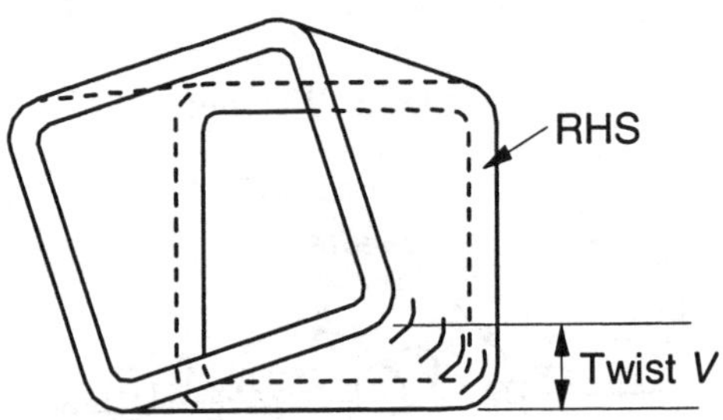


Figure 2.3 Twist of RHS and SHS

Table 2.2 Comparisons of tolerances in cross-section

| Tolerance in | Standard | | | |
|---|---|---|---|---|
| | AS 1163 | EN 10219 | ASTM A500 | G3444/G3466 |
| Outside dimension (for CHS) | +0.4 mm and -0.8 mm for $d_o \leq 50$ mm; ±0.01$d_o$ for $d_o > 50$ mm | ±0.01$d_o$ but ≥ ± 0.5mm and ≤ 10mm | ±0.005$d_o$ for $d_o \leq 48$ mm; ±0.0075$d_o$ for $d_o \geq 51$ mm | ±0.25 mm for $d_o < 50$ mm; ±0.005$d_o$ for $d_o \geq 50$ mm |
| Outside dimension (for RHS and SHS) | ±0.5 mm for $b$ or $d \leq 50$ mm; ±0.01 $b$ or ±0.01 $d$ for $b$ or $d > 50$ mm | ± 0.01 $b$ or ± 0.01 $d$ for $b$ or $d < 100$mm; ± 0.008 $b$ or ± 0.008 $d$ for 100 mm < $b$ or $d \leq 200$ mm; ± 0.006 $b$ or ± 0.006 $d$ for $b$ or $d > 200$ mm | ±0.5mm for $b$ or $d \leq 64$ mm; ±0.64 mm for 64 mm < $b$ or $d \leq 89$ mm; ±0.76 mm for 89 mm < $b$ or $d \leq 140$ mm; ±0.01 $b$ or ±0.01 $d$ for $b$ or $d > 140$ mm | ±1.5 mm for $b$ or $d \leq 100$ mm; ±0.015 $b$ or ±0.015 $d$ for $b$ or $d > 100$ mm |
| Thickness (for CHS) | ±0.1$t$ | For $d_o \leq 406.4$mm: ±0.1$t$ for $t \leq 5$ mm, ±0.5 mm for $t > 5$mm; For $d_o > 406.4$mm: ±0.1$t$ but ≤ 2 mm | ±0.1$t$ but may exceed ±0.1$t$ at the weld seam | ±0.3 mm for $t < 3$ mm; ±0.1$t$ for $3 \leq t < 12$ mm; -1.2 mm to +0.1$t$ for $t \geq 12$ mm |
| Thickness (for RHS and SHS) | ±0.1$t$ | ±0.1$t$ for $t \leq 5$ mm, ±0.5 mm for $t > 5$mm | ±0.1$t$ but may exceed ±0.1$t$ at the weld seam | ±0.3 mm for $t < 3$ mm; ±0.1$t$ for $t \geq 3$ mm |
| External corner radii (only for RHS and SHS) | < 3$t$ for all; > 1.5$t$ for RHS and SHS with perimeter equal to 50×50 or less; >1.8$t$ for RHS and SHS with perimeter greater than 50×50 | 1.5$t$ to 2.4$t$ for $t \leq 6$ mm; 2.0$t$ to 3.0$t$ for $6 < t \leq 10$ mm; 2.4$t$ to 3.6$t$ for $t > 10$ mm | < 3$t$ | < 3$t$ |

Table 2.3 Comparisons of tolerances in length, straightness, twist and mass

| Tolerance in | Standard | | | |
|---|---|---|---|---|
| | AS 1163 | EN 10219 | ASTM A500 | G3444/G3466 |
| Length | +6 mm and -0 mm for length ≤ 14 m; +10 mm and -0 mm for 14m< length ≤ 18m | +5 mm and -0 mm for length ≤ 6 m; +15 mm and -0 mm for 6 m< length ≤ 10m; + 5 mm + 1 mm/m and -0 for length > 10 m | Not specified | +nominal size and -0 mm |
| Straightness | ± $L$/500 | For CHS: ± $L$/500; For RHS and SHS: ± $L$/667 | ± $L$/480 | For CHS: ± $L$/1500 but ≤ ± 5 mm; For RHS: ± $L$/333 |
| Twist (only for RHS and SHS) | $V \leq 2$ mm + 0.5mm/m | $V \leq 2$ mm + 0.5mm/m | $V \leq 2$ mm/m for $d \leq 38.1$mm; $V \leq 1.72$ mm/m for $38.1 < d \leq 63.5$mm; $V \leq 2.09$ mm/m for $63.5 < d \leq 102$mm; $V \leq 2.42$ mm/m for $102 < d \leq 152$mm; $V \leq 2.78$ mm/m for $152 < d \leq 203$mm; $V \leq 3.11$ mm/m for $d > 203$mm | Not specified |
| Mass | ≥ 0.96$M$ | ± 0.06$M$ | Not specified | Not specified |

### 2.2.2 *Comments*

For the tolerance in CHS outside diameter, values are given in terms of a percentage of $d_o$ in EN 10219 and ASTM A500. However tolerance values are expressed in an absolute sense (i.e. mm) in AS 1163 and G3444 when $d_o$ is less than 50 mm. The tolerance given in ASTM A500 is the smallest for CHS with $d_o$ less than 50 mm. AS 1163 and EN 10219 give the largest tolerance when $d_o$ is over 50 mm.

For the tolerance in RHS/SHS outside dimensions, values are given in terms of a percentage of $b$ or $d$ in EN 10219. However tolerance values are expressed in an absolute sense (i.e. mm) in AS 1163, ASTM A500 and G3466 when $b$ or $d$ is less than 50 mm, 140 mm and 100 mm respectively. Different tolerances are given in ASTM A500 for 4 different ranges of outside dimensions, whereas only two ranges are adopted in AS 1163 and G3466. The largest tolerances are found in G3466. EN 10219 gives the smallest tolerance for RHS/SHS when the outside dimensions are over 200 mm.

For the tolerance in CHS thickness, AS 1163 and ASTM A500 have the simple rule of $0.1t$. The tolerance values in G3444 depend on the thickness of CHS. It gives a tolerance larger than 0.1t for $t < 3$ mm and a tolerance smaller than $0.1t$ for $t > 12$ mm. The tolerance values in EN 10219 depend on both the thickness and the outside diameter. It gives a tolerance smaller than $0.1t$ for a combination of $d_o$ less than 406.4 mm and $t$ larger than 5 mm.

For the tolerance in RHS/SHS thickness, AS 1163 and ASTM A500 have the simple rule of $0.1t$. G3466 gives a tolerance value larger than $0.1t$ for $t$ less than 3 mm. EN 10219 gives a tolerance value smaller than $0.1t$ when $t$ is larger than 5 mm.

For the tolerance in external corner radii, the upper value ($3t$) is the same for AS 1163, ASTM A500 and G3466. AS 1163 also specifies a lower value of $1.5t$ and $1.8t$ depending on the perimeter. EN 10219 specifies different values for tubes with different thickness.

For the tolerance in length, AS 1163, EN 10219 and G3444/G3466 all require the length to be larger than the nominal length specified. AS 1163 and EN 10219 also give an upper value depending on the length. No tolerance is specified in ASTM A500.

For the tolerance in straightness, AS 1163 and ASTM A500 do not distinguish between CHS and RHS/SHS. The tolerance values for CHS are very much the same in AS 1163, EN 10219 and ASTM A500. It is interesting to observe that the tolerance for RHS is much lower in G3466, whereas the tolerance for CHS is very high in G3444.

## 2.3 Material Properties

### *2.3.1 Mechanical Properties Specified in Manufacturing Standards*

The yield stress, tensile strength and elongations specified in AS 1163, EN 10219, ASTM A500 and G3444/G3466 are summarised in Tables 2.4, 2.5 and 2.6. It can be seen that the yield stress ranges from 228 to 460 MPa (N/mm$^2$) while the tensile strength rages from 310 to 720 MPa. The ratio ($f_u/f_y$) ranges from 1.11 to 1.92. The elongation is between 14% and 25% with slightly different gauge lengths defined in various standards. The symbol $S_o$ in Table 2.6 is the cross-sectional area of the undeformed tensile coupon.

AS 1163 also states that prior to tensile or impact tests, the test series shall be aged by heating to a temperature between 150 and 200 degrees Celsius for a period not less than 15 minutes. By comparison, ASTM A500 states that the minimum elongation values apply only to testing performed prior to shipment because of the possibility of strain ageing. Tests performed after shipment may show values above those stated.

Table 2.4 Minimum values of yield stress, tensile strength and tensile to yield ratio for CHS

| Standard | Grade | Minimum yield stress ($f_y$) in N/mm$^2$ | | Minimum tensile strength ($f_u$) in N/mm$^2$ | $f_u/f_y$ |
|---|---|---|---|---|---|
| AS 1163 | C250, C250L0 | 250 | | 320 | 1.28 |
| | C350, C350L0 | 350 | | 430 | 1.23 |
| | C450, C450L0 | 450 | | 500 | 1.11 |
| EN 10219 | | $t \leq 16$ mm | 16 mm $< t \leq 40$ mm | $t \leq 40$ mm | |
| | S275 NH S275 NLH | 275 | 265 | 370 – 510 | 1.35 to 1.92 |
| | S355 NH S355 NLH | 355 | 345 | 470 – 630 | 1.33 to 1.83 |
| | S460 NH S460 NLH | 460 | 440 | 550 – 720 | 1.20 to 1.63 |
| ASTM A500 | A | 228 | | 310 | 1.36 |
| | B | 290 | | 400 | 1.38 |
| | C | 317 | | 427 | 1.35 |
| | D | 248 | | 400 | 1.61 |
| G3444 | STK41 | 235 | | 402 | 1.71 |
| | STK50 | 314 | | 490 | 1.56 |
| | STK51 | 353 | | 500 | 1.42 |
| | STK55 | 392 | | 539 | 1.38 |

Table 2.5 Minimum values of yield stress, tensile strength and tensile to yield ratio for RHS/SHS

| Standard | Grade | Minimum yield stress ($f_y$) in N/mm$^2$ | | Minimum tensile strength ($f_u$) in N/mm$^2$ | $f_u/f_y$ |
|---|---|---|---|---|---|
| AS 1163 | C250, C250L0 | 250 | | 320 | 1.28 |
| | C350, C350L0 | 350 | | 430 | 1.23 |
| | C450, C450L0 | 450 | | 500 | 1.11 |
| EN 10219 | | $t \leq 16$ mm | 16 mm $< t \leq 24$ mm | $t \leq 24$ mm | |
| | S275 NH S275 NLH | 275 | 265 | 370 – 510 | 1.35 to 1.92 |
| | S355 NH S355 NLH | 355 | 345 | 470 – 630 | 1.33 to 1.83 |
| | S460 NH S460 NLH | 460 | 440 | 550 – 720 | 1.20 to 1.63 |
| ASTM A500 | A | 269 | | 310 | 1.15 |
| | B | 317 | | 400 | 1.26 |
| | C | 345 | | 427 | 1.24 |
| | D | 248 | | 400 | 1.61 |
| G3466 | STKR41 | 245 | | 402 | 1.64 |
| | STKR50 | 324 | | 490 | 1.51 |

It can be noted that only AS 1163 and EN 10219 include the higher yield stress grade of 440 – 460 N/mm$^2$.

Table 2.6 Minimum values of elongation

| Standard | Grade | Elongation (%) for CHS | Elongation (%) for RHS/SHS | Gauge length used |
|---|---|---|---|---|
| AS 1163 | C250, C250L0 | 22 | 18 | $5.65 \cdot \sqrt{S_o}$ |
| | C350, C350L0 | 20 | 16 | |
| | C450, C450L0 | 16 | 14 | |
| EN 10219 | S275 NH S275 NLH | 24 | 24 | $5.65 \cdot \sqrt{S_o}$ |
| | S355 NH S355 NLH | 22 | 22 | |
| | S460 NH S460 NLH | 17 | 17 | |
| ASTM A500 | A | 25 | 25 | 50.8 mm |
| | B | 23 | 23 | |
| | C | 21 | 21 | |
| | D | 23 | 23 | |
| G3444 | STK41 | 23 | | 50 mm |
| | STK50 | 23 | | |
| | STK51 | 15 | | |
| | STK55 | 20 | | |
| G3466 | STKR41 | | 23 | 50 mm |
| | STKR50 | | 23 | |

The C450 sections in AS 1163 may have a lower elongation due to enhanced yield from a cold-forming process. By comparison EN 10219 S460 sections have a higher elongation (17%).

### *2.3.2 Variation of Yield Stress around a Section*

The mechanical properties (Young's modulus, yield stress, tensile strength and elongation) of cold-formed tubular sections can be obtained from tensile coupon tests in accordance with certain material testing standards, such as AS 1391 (Standards Australia 1991) in Australia and ASTM A370 (1995) in the USA. Cold-formed tubular sections tend to have rounded stress-strain curves. The 0.2% proof stress is commonly used to define the yield stress. A schematic view of stress-strain curves for the adjacent faces, the opposite face and the corner is given in Figure 2.4.

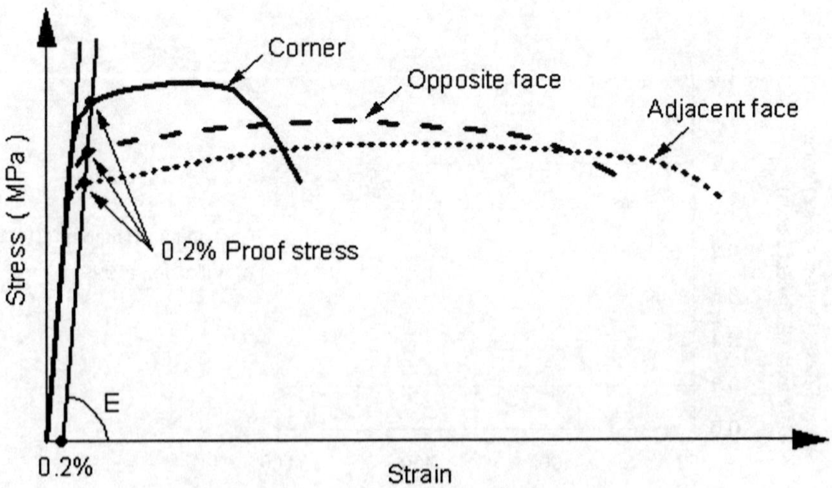


Figure 2.4 Schematic view of stress-strain curves

Variation was found in material properties around a cold-formed RHS (Key and Hancock 1985, Key et al 1988, Zhao and Hancock 1991, 1992, Zhao and Grzebieta 2002, Wilkinson 1999). The yield stress and tensile strength in the opposite face were found to be higher than those in the adjacent faces because the opposite face is subject to more bending in the forming process. The yield stress and tensile strength in the corners were found to be even higher because more cold-working exists in the corners. The comparisons are illustrated in Figures 2.5 and 2.6. The average yield stress of flat faces (i.e. average of yield stress in opposite face and those in adjacent faces) is chosen as the horizontal axis in both figures. The average ratios are summarized in Table 2.7.

Table 2.7 Summary of comparisons of mechanical properties

| Comparison | Ratio on average | Coefficient of Variation |
|---|---|---|
| Yield stress (opposite face to adjacent faces) | 1.09 | 0.0378 |
| Tensile strength (opposite face to adjacent faces) | 1.06 | 0.0320 |
| Yield stress (corners to average flat faces) | 1.22 | 0.0727 |
| Tensile strength (corners to average flat faces) | 1.16 | 0.0575 |

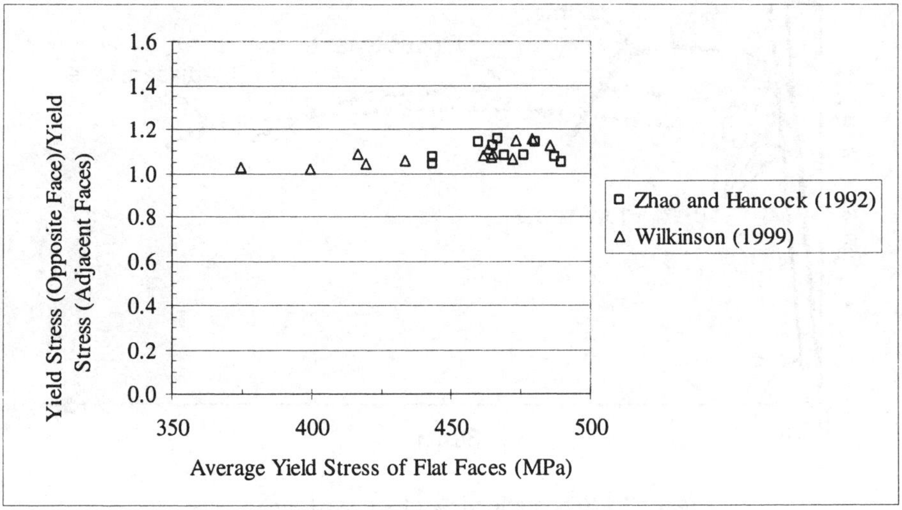


(a)

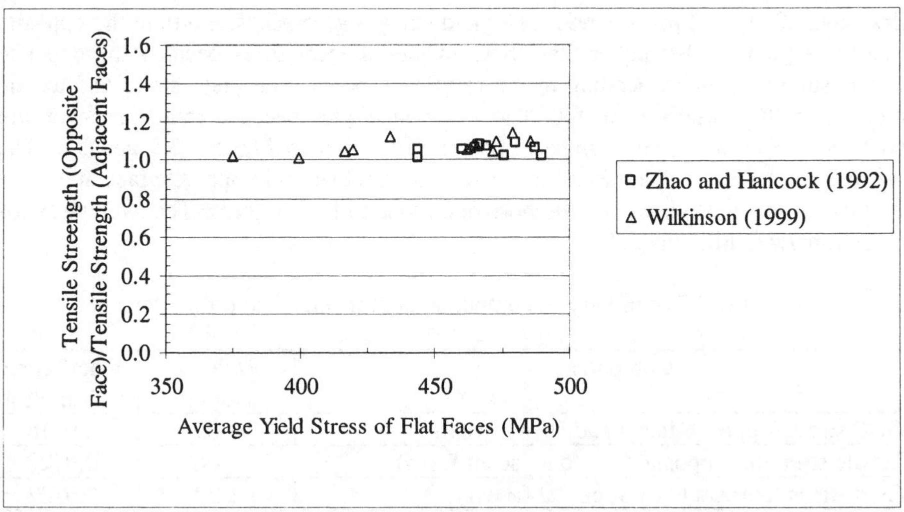


(b)

Figure 2.5 Comparison of opposite face and adjacent faces (a) yield stress, (b) tensile strength

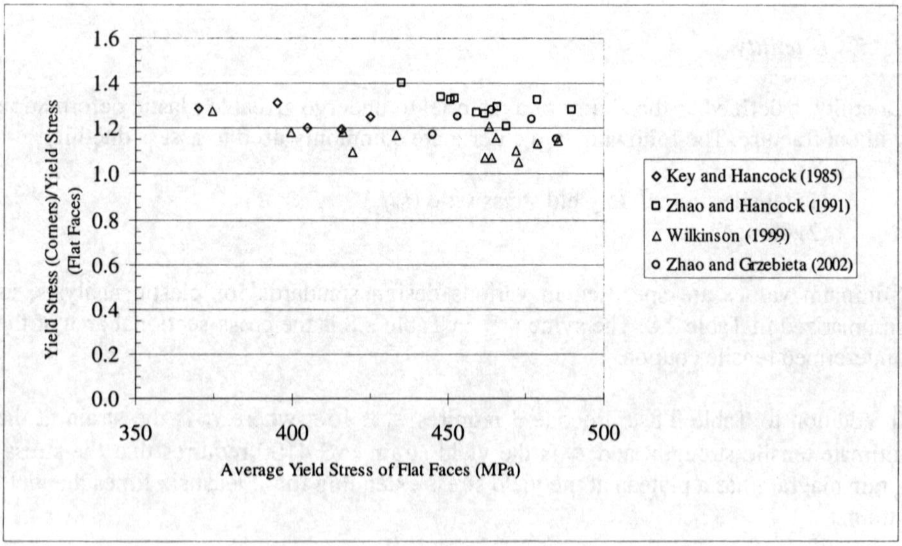


(a)

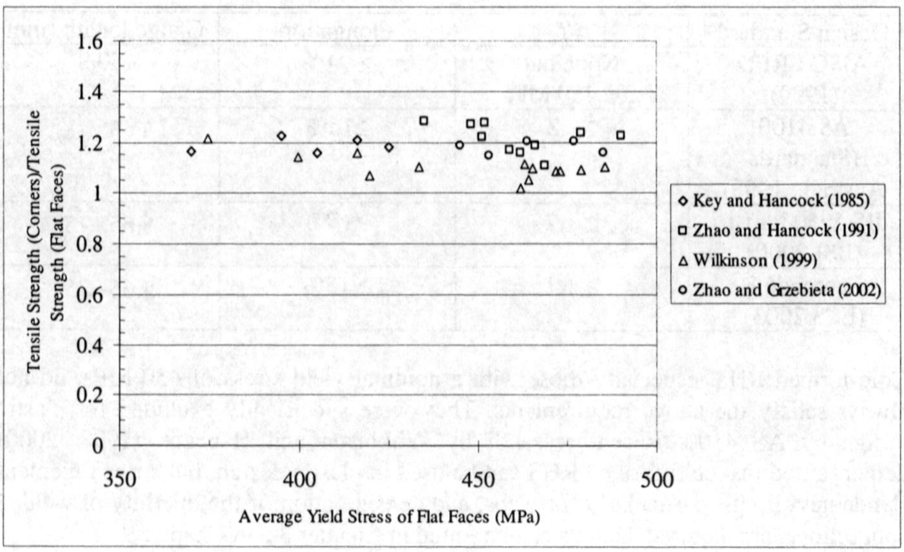


(b)

Figure 2.6 Comparison of corners and flat faces (a) yield stress, (b) tensile strength

### 2.3.3 *Ductility*

Ductility is defined as the ability of a material to undergo sizeable plastic deformation without fracture. The following two criteria are commonly used to assess ductility.

(1) tensile strength to yield stress ratio ($f_u/f_y$)
(2) elongation

Minimum values are specified in various design standards for plastic analysis, as summarized in Table 2.8. The symbol $S_o$ in Table 2.8 is the cross-sectional area of the undeformed tensile coupon.

In addition to Table 2.8, Eurocode 3 requires $\varepsilon_u \geq 15\varepsilon_y$ where $\varepsilon_u$ is the strain at the ultimate tensile strength, and $\varepsilon_y$ is the yield strain. AS 4100 requires that the stress-strain diagram has a plateau at the yield stress extending for at least six times the yield strain.

Table 2.8 Material property requirement

| Design Standard | $f_u/f_y$ | Elongation | Gauge length (mm) |
|---|---|---|---|
| AISC LRFD (1999) | None but $f_y < 450$ MPa | $\geq 21\%$ | 50.8 |
| AS 4100 (Standards Australia 1998) | $\geq 1.2$ | $\geq 15\%$ | $5.65 \cdot \sqrt{S_o}$ |
| BS 5950 Part 1 (BSI 2000) | $\geq 1.2$ | $\geq 15\%$ | $5.65 \cdot \sqrt{S_o}$ |
| Eurocode 3 (EC3 2003) | $\geq 1.1$ | $\geq 15\%$ | $5.65 \cdot \sqrt{S_o}$ |

Cold-formed RHS, especially those with a nominal yield stress of 450 MPa, do not always satisfy the above requirements. They were specifically excluded for plastic design by AS 4100. Recent research by Wilkinson and Hancock (1998, 2000) demonstrated that cold-formed RHS can be used in plastic design, but stricter element slenderness limits, particularly for webs, and consideration of the ductility of welded connections, are required. Details can be found in Chapter 3 and Chapter 9.

### 2.3.4 *Residual Stress*

Residual stresses are generated in steel tubes during the manufacturing process. It was found in Key and Hancock (1985) that for cold-formed C350 RHS: "in general the average residual stress on each face is equal to approximately half the actual yield stress, compressive on the inside and tensile on the outside". Similar results were found for CHS in the longitudinal direction (Jiao and Zhao 2002). The residual stress along the CHS transverse direction was found to be about 40% of those along the longitudinal direction (Jiao and Zhao 2002). Typical values are shown in Table 2.9.

The through-thickness variation of residual stress in both the longitudinal and transverse directions at the centre of one face of a 254×254×6.3 SHS was measured

using a spark erosion layering technique (Key and Hancock 1993). The three stages, namely Panel Removal, Small Block Removal and Small Block Spark Erosion Layering, involved in the spark erosion technique are shown in Figure 2.7 together with the calculated residual stress components. It seems that the released residual stress from Small Block Removal was very small compared with those in the other two stages. The residual stresses in the Panel Removal stage show a linear distribution across the thickness. The residual stresses in the final stage tend to vary linearly from the surface to the centre of the plate.

Table 2.9 Typical values of residual stress in cold-formed sections

| Section Type | Longitudinal or Transverse | Inside or Outside Surface | As Percentage of Yield Stress | Compression or Tension | Reference |
|---|---|---|---|---|---|
| RHS | Longitudinal | Outside | 50 | Tension | Key and Hancock (1985) |
| RHS | Longitudinal | Inside | 50 | Compression | |
| CHS | Longitudinal | Outside | 64 | Tension | Jiao and Zhao (2002) |
| CHS | Longitudinal | Inside | 48 | Compression | |
| CHS | Transverse | Outside | 17 | Tension | |
| CHS | Transverse | Inside | 25 | Compression | |

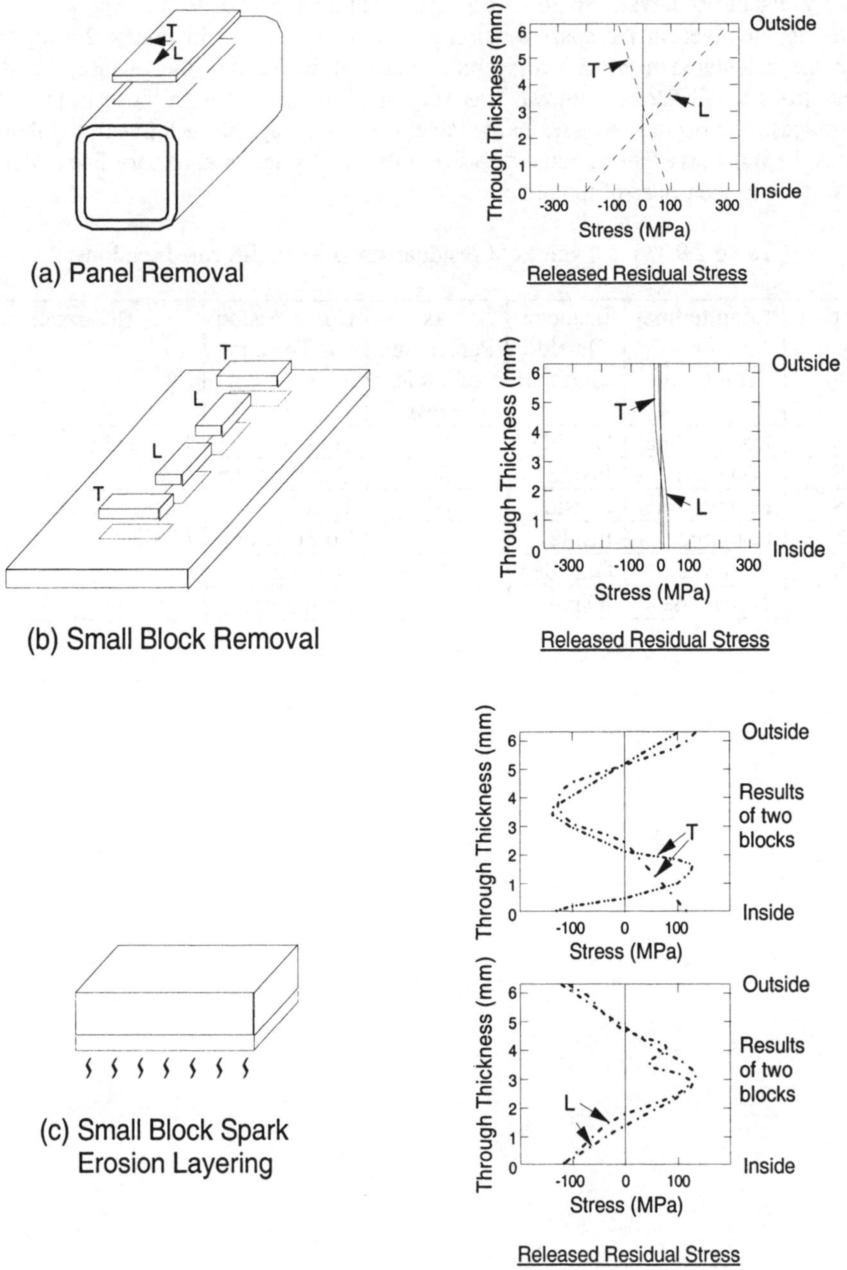


Figure 2.7 Measured through-thickness residual stress (Key and Hancock 1993)

### *2.3.5 Fracture Toughness*

Fracture toughness is an important design parameter for cold-formed tubes under dynamic loading. There is an increasing awareness of material notch ductility after the Northridge and Kobe earthquakes. Fracture toughness can be measured using the Charpy V-Notch impact test. Most manufacturing standards specify the position and orientation of the specimen to be cut from a section for Charpy V-Notch (CVN) test, i.e. the flat face location in the longitudinal direction (Dagg et al 1989). Recently an extensive study was conducted by Kosteski et al (2005) on fracture toughness of cold-formed hollow sections. The study included specimens taken from different orientations (longitudinal and transverse), cross-section location (flat, corner and weld seam) and face exposure (interior face and exterior face of the tube). It was found by Kosteski et al (2005) that the flat face transverse CVN coupon generally results in lower fracture toughness than the flat face longitudinal CVN coupon. Improved fracture toughness was found for cold-formed tubes after induction heating (Tanaka et al 1996).

## 2.4 Special Characteristics

Some special characteristics associated with cold-formed tubes are listed below. They are addressed in the book later.

- Rounded stress-strain material behaviour and variation of yield stress around the section.
- Larger residual stresses that may result in a lower column curve.
- Lateral buckling and torsion are not as major a concern for steel hollow sections compared to open sections.
- Web crippling of RHS due to the external corner radii that introduce load eccentrically to the webs although RHS have two webs to compensate to some degree.
- Interaction of web local buckling and flange local bucking in bending.
- Oversized fillet weld for thin-walled tubes, which may increase the strength of fillet-welded connections
- Weld defects in welded thin-walled tubes and their impact on fatigue strength.
- Corner cracking in thick-walled cold-formed tubes.
- Challenge for plastic design because of lower ductility.

## 2.5 Limit States Design

As mentioned in Chapter 1, design examples will be given in this book in accordance with AS 4100 and BS 5950 Part 1. Both standards adopted the limit states design approach. A brief description of the limit states design is given in this section. The differences in AS 4100 and BS 5950 Part 1 are pointed out.

Limit states design is a design method in which the performance of a structure is checked against various limiting conditions at appropriate load levels. The limiting conditions to be checked in structural steel design are ultimate limit state and

serviceability limit state. Ultimate limit states are those states concerning safety, such as exceeding of load-carrying capacity, overturning, sliding and fracture due to fatigue or other causes. Serviceability limit states are those in which the behaviour of the structure is unsatisfactory, and include excessive deflection, excessive vibration and excessive permanent deformation. It is the ultimate limit state (or strength limit state) that is dealt with in the examples of this book.

For the strength limit state design, the structure is deemed to be satisfactory if its design load effect does not exceed its design resistance. In AS 4100 this criterion is described as:

$$S^* \leq \phi \cdot R_u \tag{2.1}$$

where $S^*$ is the design action effect (e.g. design bending moments, shear forces and axial forces), $\phi$ is the capacity factor and $R_u$ is the nominal capacity.

Load factors are used in determining the design action effects ($S^*$) as specified in AS/NZS 1170.0 (Standards Australia 2002). For example, a load factor of 1.2 is given to the dead load and a load factor of 1.5 is given to the live load for static design. When checking structures or elements of structures for fatigue performance under repeated in-service cyclic actions, the level of repeated loading to be used shall be the actual load level expected for the design situation. The capacity factor $\phi$ takes into account the fact that the actual strength of a member may be less than anticipated due to variability of material properties, fabrications and design models. Reliability analysis is often used to calibrate the capacity factors (Ravindra and Galambos 1978). The values of the capacity factor are given in Table 3.4 of AS 4100. For member design the capacity factor is 0.9. For connection design the capacity factor varies from 0.6 to 0.9.

In BS 5950 Part 1 the strength limit state criterion is expressed in a slightly different format (Trahair et al 2001) as:

$$\sum \gamma_f \times (\text{effect of specified loads}) \leq (\text{specified resistance} / \gamma_m) \tag{2.2}$$

where $\gamma_f$ is the partial factor for loads and $\gamma_m$ is the partial factor for material strength.

For example, a load partial factor $\gamma_f$ of 1.2 is given to the dead load and a load partial factor of 1.6 is given to the live load for static design. For fatigue design a $\gamma_f$ factor of 1.0 should be used. In BS 5950 Part 1 the material partial factor $\gamma_m$ is incorporated in the recommended design strengths. For structural steel the material factor is taken as 1.0 applied to the yield stress or 1.2 applied to the tensile strength. Different values are used for bolts and welds.

It can be seen that AS 4100 and BS 5950 Part 1 adopt the same load factor for dead load whereas they adopt slightly different load factors for live load, i.e. 1.5 in AS 4100 and 1.6 in BS 5950 Part 1. AS 4100 applies the capacity factor to the nominal strength while BS 5950 Part 1 incorporates the material partial factor in the recommended design strengths. This should be kept in mind when comparing design capacities obtained from AS 4100 and BS 5950 Part 1 in Chapters 3 to 8.

## 2.6 References


1. AISC (1999), Load and Resistance Factor Design Specification for Structural Steel Buildings, American Institute of Steel Construction, Chicago, Illinois, USA
2. ASTM (1993), Standards Specification for Cold-Formed Welded and Seamless Carbon Steel Structural Tubing in Rounds and Shapes, American Society for Testing Materials ASTM A500, USA
3. ASTM (1995), Standard Test Methods and Definition for Mechanical Testing of Steel Products, ASTM A370, USA
4. BSI (2000), Structural use of Steelwork in Building, BS 5950, Part 1, British Standard Institution, London, UK
5. Dagg, H.M., Davis, K. and Hicks, J.W. (1989), Charpy Impact Tests on Cold-Formed RHS Manufactured from Continuous Cast Fully Killed Steel, *Proceedings*, Pacific Structural Steel Conference, Brisbane, Australia
6. EC3 (2003), Eurocode 3: Design of Steel Structures – Part 1.1: General Rules and Rules for Buildings, prEN 1993-1-1:2003, November 2003, European Committee for Standardization, Brussels, Belgium
7. ENV (1992), European Committee for Standardization, European Pre-Standard ENV 10219, Cold-Formed Welded Structural Hollow Sections of Non-Alloyed and Fine Grained Steels, Part 1 Technical Delivery condition, Part 2 Tolerances, Dimensions and Section Properties, British Standards Institution, London, UK
8. Jiao, H. and Zhao, X.L. (2002), Imperfection, Residual Stress and Yield Slenderness Limit of Very High Strength (VHS) Circular Steel Tubes, *Journal of Constructional Steel Research*, 59(2), pp 233-249
9. JIS (1988a), Carbon Steel Tubes for General Structural Purposes, Japanese Industrial Standard, G3444, Tokyo, Japan
10. JIS (1988b), Carbon Steel Square Pipe for General Structural Purposes, Japanese Industrial Standard, G3466, Tokyo, Japan
11. Key, P.W. and Hancock, G.J. (1985), An Experimental Investigation of the Column Behaviour of Cold-Formed Square Hollow Sections, *Research Report R493*, School of Civil and Mining Engineering, The University of Sydney, Sydney, Australia
12. Key, P.W., Hasan, S.W. and Hancock, G.J. (1988), Column Behaviour of Cold-Formed Hollow Sections, *Journal of Structural Engineering*, ASCE, 114(2), pp 390-407
13. Key, P.W. and Hancock, G.J. (1993), A Theoretical Investigation of the Column Behaviour of Cold-Formed Square Hollow Sections, *Thin-Walled Structures*, 16(1-4), pp 31-64
14. Kosteski, N., Packer, J.A. and Puthli, R.S. (2005), Notch Toughness of Internationally Produced Hollow Sections, *Journal of Structural Engineering,* ASCE, 131(2), pp 279-286
15. Ravindra, M.K. and Galambos, T.V. (1978), Load and Resistance Factor Design for Steel, *Journal of the Structural Division,* ASCE, 104(ST9), pp 1337-1353
16. Standards Australia (1991), Structural Steel Hollow Sections, Australian Standard AS 1163, Standards Australia, Sydney, Australia
17. Standards Australia (1991), Methods for Tensile Testing of Metals, Australian Standard AS 1391, Standards Australia, Sydney, Australia
18. Standards Australia (1998), Steel Structures, Australian Standard AS 4100, Standards Australia, Sydney, Australia

19. Standards Australia (2002), Structural Design Actions Part 0: General Principles, Australian/New Zealand Standard AS/NZS 1170.0, Standards Australia, Sydney, Australia
20. Tanaka, T., Tabuchi, M., Murayama, M., Furumi, K., Morita, T., Usami, K., Matsubara, Y. (1996), Experimental Study on End-plate to SHS Column Connections Reinforced by Increasing Wall Thickness with One Side Bolts, In: *Tubular Structures VII*, Farkas, J. and Jármai, K. (eds), Balkema: Rotterdam, The Netherlands, pp 253-260
21. Trahair, N.S., Bradford, M.A. and Nethercot, D.A. (2001), *The Behaviour and Design of Steel Structures to BS5950*, Spon Press, London, UK
22. Wilkinson T. (1999), The Plastic Behaviour of Cold-Formed Rectangular Hollow Sections, *PhD Thesis*, The University of Sydney, Sydney, Australia
23. Wilkinson T. and Hancock, G.J. (1998), Tests to Examine the Compact Web Slenderness of Cold-Formed RHS, *Journal of Structural Engineering*, ASCE, 124(10), pp 1166-1174
24. Wilkinson T. and Hancock, G.J. (2000), Tests to Examine Plastic Behaviour of Knee Joints in Cold-Formed RHS, *Journal of Structural Engineering*, ASCE, 126(3), pp 297-305
25. Zhao, X.L. and Hancock, G.J. (1991), Tests to Determine Plate Slenderness Limits for Cold-Formed Rectangular Hollow Sections of Grade C450, *Steel Construction*, Journal of the Australian Institute of Steel Construction, 25(4), pp 2-16
26. Zhao, X.L. and Hancock, G.J. (1992), Tests to Determine the Reliability of Stub Columns of DuraGal RHS, *Investigation Report S916*, Centre for Advanced Structural Engineering, The University of Sydney, Sydney, Australia
27. Zhao, X.L. and Grzebieta, R.H. (2002), Strength and Ductility of Concrete Filled Double Skin (SHS Inner and SHS Outer) Tubes, *Thin-Walled Structures*, 40(2), pp 199-213

# Chapter 3: Members Subjected to Bending

## 3.1 Introduction

The most common case in which a hollow section experiences bending would be as a horizontal beam supporting transverse gravity loads. However hollow sections are subjected to bending in other applications, such as:

- Low rise portal frame structures in which both the beams and columns experience predominantly bending loads.
- Rigid jointed truss structures with loading along the chord members.
- Transportation systems such as trailers or rollover cages.

Additional examples and photographs of hollow sections in bending applications are given in Chapter 1.

When a transverse load ($P$) is applied to a steel beam, as shown in Figure 3.1, there is corresponding deformation, which includes curvature ($\kappa$) (the inverse of the radius of curvature of the bent profile), induced in the beam. Internal forces, such as bending moments ($M$), occur within the beam.

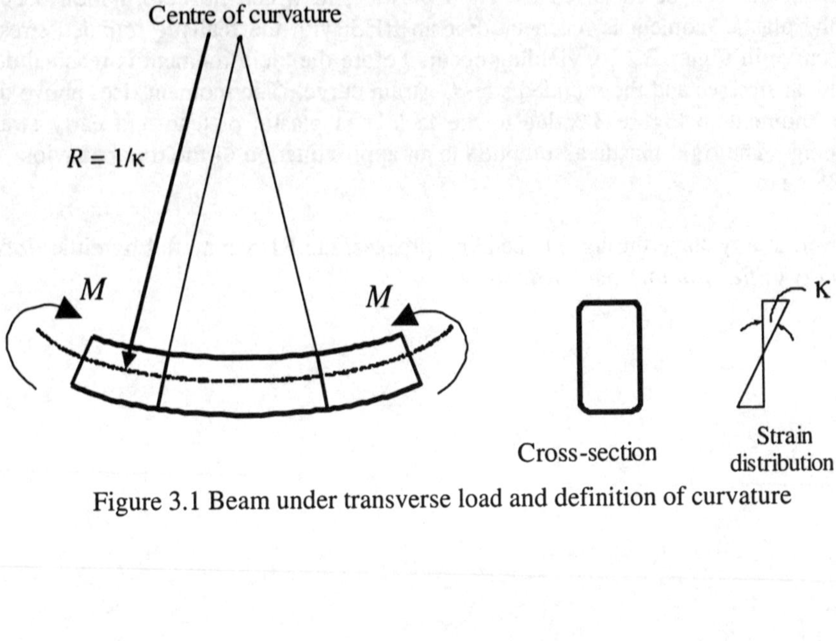


Figure 3.1 Beam under transverse load and definition of curvature

For the RHS in bending, the distribution of strain across the section is assumed linear according to engineering bending theory regardless of the stress state, and the value of the strain at the extreme fibres is proportional to the curvature. Figure 3.2 indicates how the stress distribution changes with increasing levels of curvature for an RHS with either the idealised elastic - plastic - strain hardening material properties, or the gradually yielding stress - strain behaviour of cold-formed steel. Initially, in the elastic range, the stress distribution is linear. As the curvature increases, the extreme fibres reach the yield stress at the yield moment ($M_y$), where $M_y = f_y Z$ and $Z$ is the elastic section modulus. At larger curvatures and strains, yielding spreads inwards toward the neutral axis. For the elastic - plastic - strain hardening material, the section yields almost completely and is fully plastic at high values of curvature (theoretically full plasticity can only occur at infinite curvature). The theoretical moment at which full yielding occurs is termed the plastic moment ($M_p$), where $M_p = f_y S$ and $S$ is the plastic section modulus. Strain hardening is initiated at high curvatures, and the stress can exceed the yield stress and the moment can exceed the plastic moment. In the case of cold-formed RHS with the stress-strain characteristics of Figure 2.4, there is generally no significant plastic plateau as strain hardening occurs immediately after yielding, and the stress increases beyond $f_y$ at low values of curvature compared to the case of an elastic – perfectly plastic material.

The resulting idealised moment-curvature relationships of the cross-section are shown in Figure 3.3. For the idealised case, the curve includes a linear range and a transition from the yield moment to the plastic moment. Once the cross-section is fully plastic, increases in curvature can occur without a corresponding moment increase. Not only does the moment reach $M_p$ but the beam maintains $M_p$ as the curvature increases. The increasing curvature at constant moment $M_p$ is termed a *plastic hinge* and demonstrates the ductility of the steel beam. The moment may rise above the plastic moment due to strain hardening, but the increase in moment is sometimes ignored. The behaviour can be idealised as "rigid plastic", in which no deformation occurs until the plastic moment is reached. For an RHS with the realistic rounded stress-strain curve in Figure 3.2(b), yielding occurs before the yield moment is reached due to residual stresses and the rounded stress - strain curve. The moment rises above the plastic moment in Figure 3.3 due to the lack of a plastic plateau and early strain hardening. The rigid-plastic assumption is an approximation of the true behaviour of an RHS beam.

However, at any stage during this bending process, the RHS may fail by either *local instability* or *flexural torsional instability*.

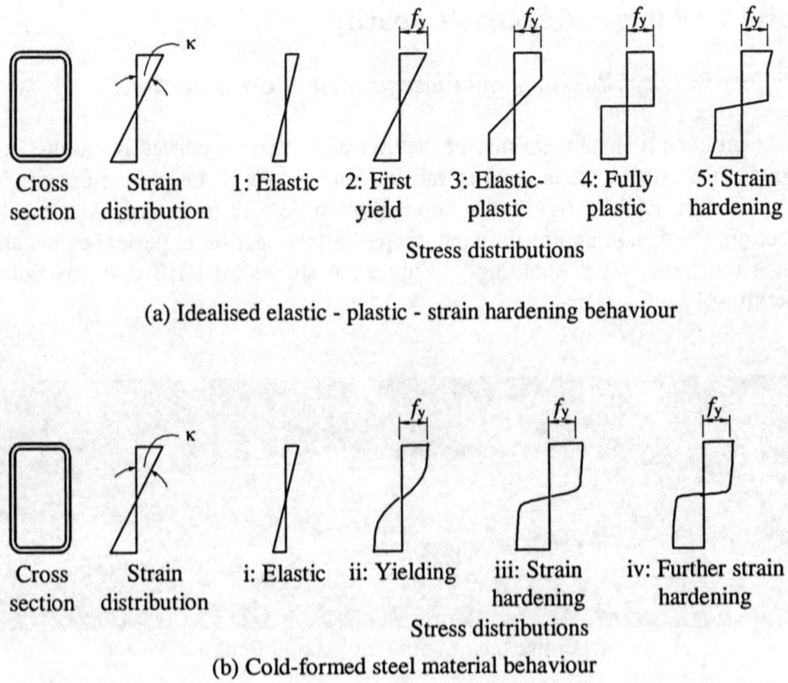


Figure 3.2 Strain and stress distributions in hollow sections in bending

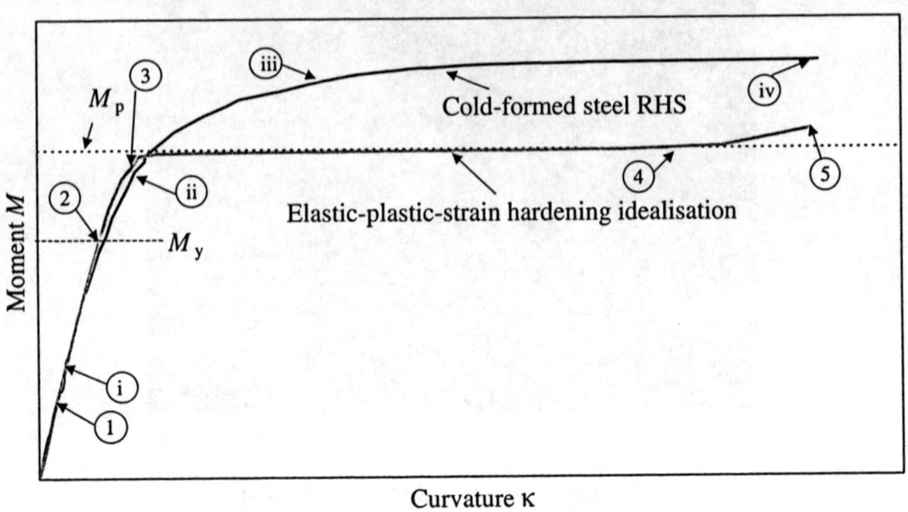


Figure 3.3 Moment-curvature relationship for a hollow section in bending

## 3.2 Local Buckling and Section Capacity

### *3.2.1 Failure by Local Buckling and Classification of Cross-Sections*

A steel beam cannot sustain infinite curvature, and at some curvature failure occurs. A common mode of failure is local instability (buckling) of the plate elements in the section, although material fracture is another possible failure mode. At a particular cross-section, the flange and/or the web of the hollow section experiences instability, and hence the term "local buckling". Figure 3.4 shows an RHS that has failed by local buckling.

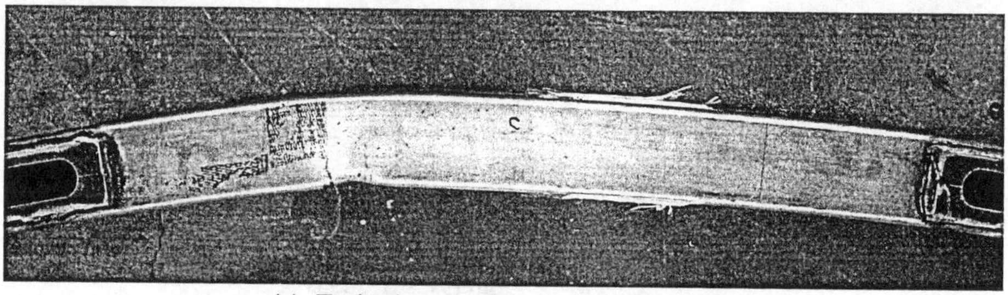

(a) Entire beam with single local buckle

(b) Magnified region of the local buckle

Figure 3.4 RHS beam failed by local buckling (Jouaux 2004)

Some beams may fail before reaching the yield moment or the plastic moment. If the beam can reach the plastic moment, the *rotation capacity* ($R$) is a measure of how much the plastic hinge can rotate before failure occurs. To calculate $R$, the moment-

curvature graph is normalised with respect to the plastic moment and plastic curvature ($\kappa_p = M_p/EI$), (where $E$ is the Young's modulus of elasticity, or elastic modulus, and $I$ is the second moment of area of the section). Assuming buckling occurs after the moment increases above $M_p$, then the moment drops below $M_p$ at some curvature ($\kappa_1$). The rotation capacity is commonly defined as $R = \kappa_1/\kappa_p - 1$, where $\kappa_p = M_p/EI$.

Sections are classified into groups depending on their behaviour under bending, (their rotation capacity and maximum moment, $M_{max}$), as illustrated in Figure 3.5. Some design standards define four classes of sections, while other steel standards define only three separate classes.

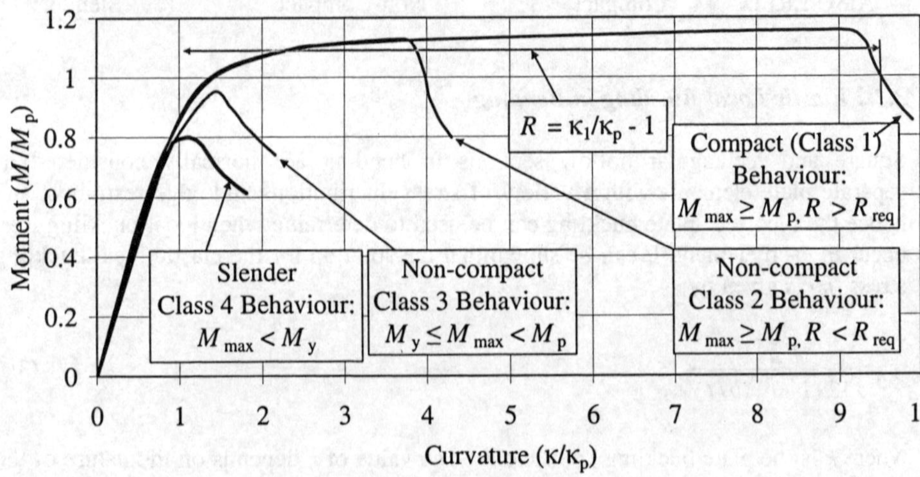


Figure 3.5 Moment-curvature behaviour of different types of steel sections

1) *Class 1* sections can attain the plastic moment and have plastic rotation capacity sufficient for plastic design. Such sections are sometimes referred to as *plastic* sections (BS 5950), or *compact* sections (AS 4100, AISC LRFD).
2) *Class 2* sections can develop the plastic moment but have limited rotation capacity and are considered unsuitable for plastic hinge formation. Class 2 sections may be known as *compact* sections (BS 5950, CSA-S16.1) or *compact elastic* (Galambos 1976) sections. Confusion may arise with the dual use of the term "compact" for Class 1 in AISC LRFD and AS 4100, and Class 2 in CSA-S16.1.
3) *Class 3* sections can reach the yield moment, but cannot reach the plastic moment due to local buckling. Such sections are sometimes called *semi-compact* (BS 5950), or *non-compact* (CSA-S16.1).
4) *Class 4* sections cannot reach the yield moment due to local buckling. They are also known as *slender* sections in all standards.

Some specifications, such as AS 4100 and AISC LRFD, group together Class 2 and Class 3 sections, into one single class, commonly referred to as *non-compact*. Under

the AS 4100 and AISC LRFD definition, "non-compact" sections have a moment capacity exceeding the yield moment, and up to and including the plastic moment, but cannot sustain the plastic moment for suitably large rotations. The moment capacity for such sections varies linearly with slenderness from the yield moment to the plastic moment. Table 3.1 summarises the terminology in some of the more widely known steel design codes.

Table 3.1 Section classification in various design standards

| Specification | Section classification | | | |
|---|---|---|---|---|
| Eurocode 3 | Class 1 | Class 2 | Class 3 | Class 4 |
| BS 5950 | Plastic | Compact | Semi-Compact | Slender |
| AS 4100 | Compact | Non-Compact | | Slender |
| AISC LRFD | Compact | Non-Compact | | Slender |

### 3.2.2 Elastic Local Buckling in Bending

Square and rectangular hollow sections in bending are normally considered as separate plate elements with a variety of stress distributions and edge restraints, and hence the theory of plate buckling can be used to determine when local buckling may occur in each element. It can be shown that the solution for the elastic local buckling stress ($f_o$) is given by:

$$f_o = \frac{k \pi^2 E}{12(1-\nu^2)(b/t)^2} \tag{3.1}$$

where $k$ is the plate buckling coefficient. The value of $k$ depends on the nature of the stress distribution across the plate and the support conditions of the plate as shown in Figure 3.6.

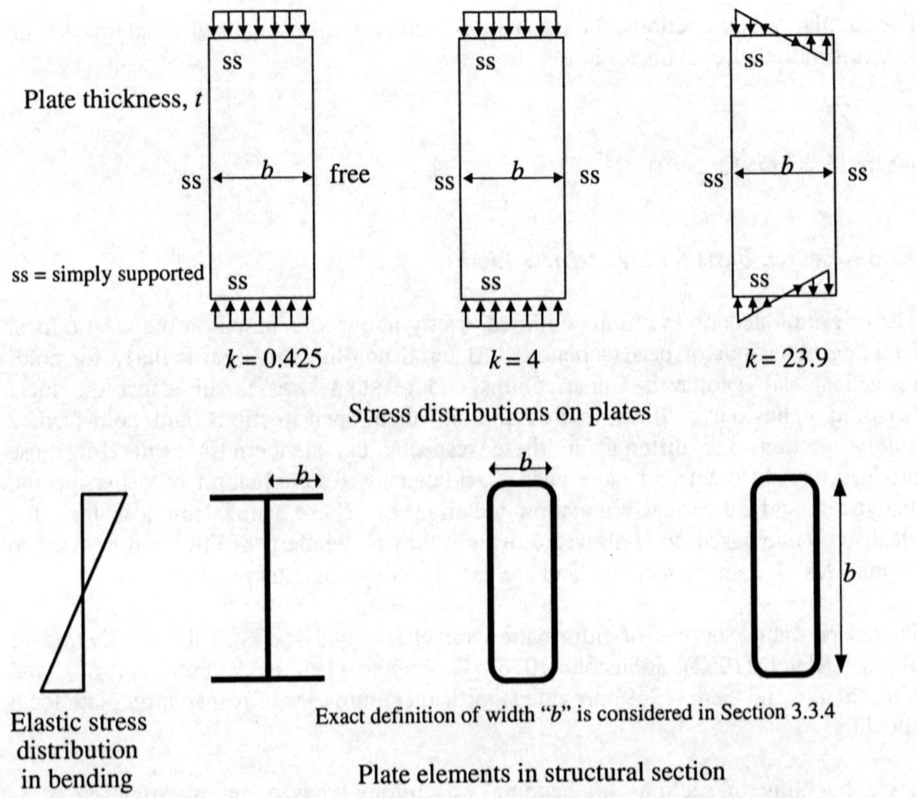


Figure 3.6 Stress distributions on a plate

Hence, to prevent a plate from buckling before it reaches its yield stress (ie to avoid elastic local buckling), then $f_y \le f_o$, or

$$f_y \le \frac{H^2}{(b/t)^2} \tag{3.2}$$

where $H^2 = k\pi^2 E/\{12(1-\nu^2)\}$.

Equation (3.2) can be written in terms of a *slenderness limit* for $b/t$ for some value $H$ to prevent elastic local buckling:

$$\frac{b}{t} \le \frac{H}{\sqrt{f_y}} \tag{3.3}$$

For the case of an RHS flange, $k = 4.0$, $E = 200000$ MPa, $\nu = 0.3$, then $H = 850$. For $f_y = 275$ MPa, the idealised limit for $b/t = 51.3$.

The appropriate value of the constant $H$ depends on the stress distribution and support conditions as well as allowances for level of residual stress and imperfections in the member.

For circular hollow sections, the rationale is similar but the idealised slenderness limit in terms of the outer diameter is

$$\frac{d_o}{t} \leq \frac{H'}{f_y} \quad (3.4)$$

where $H' = 2E/\sqrt{\{3(1-\nu^2)\}}$

### 3.2.3 *Research Basis for Slenderness Limits*

The origin of slenderness limits outlined briefly above was based on the elastic local buckling behaviour of perfect plates. Material non-linearity (particularly for cold-formed steels), geometric imperfections and residual stresses all affect the local buckling behaviour. Since traditional hot-rolled open sections and cold-formed hollow sections are different in these respects, the slenderness limits for these products are different and have been calculated by a combination of experimental, analytical, and numerical simulation techniques. These values are less than the idealised value based on $H$ above. Lower values of slenderness limit can be used to define Class 1, 2 and 3 sections, and are based on experimental results.

There are many sources of information on elastic and inelastic local buckling of plates. Bleich (1952), Johnston (1976), Ostapenko (1983), Galambos (1968), and Timoshenko and Gere (1969) provide significant summaries of research on plate local buckling.

Local buckling of sections on bending was initially based on investigation of I-sections, with much work on the flanges performed in the 1960s. Ueda and Tall (1967), Lay (1965), Lukey and Adams (1969) and Kato (1965) are some of the many investigations. The basis of I-section web slenderness limits comes from investigations such as Haaijer and Thurlimann (1958) and Dawe and Kulak (1984a, 1984b, 1986).

Tests on hollow sections include the investigations of Korol and Hudoba (1972), Hasan and Hancock (1988), Zhao and Hancock (1991), Corona and Vaze (1996), Stranghöner (1995), Wilkinson and Hancock (1998) and Wilkinson (2003). Tests on fabricated CHS are available from Sherman (1986), and for cold-formed CHS from Elchalakani et al (2002a, 2002b).

### 3.2.4 *Slenderness Limits in Current Specifications*

It was shown in Section 3.2.1 that local buckling behaviour can be classified according to the element slenderness of the flange or web of the cross-section (or the diameter to thickness ratio for a CHS). The definition of slenderness varies slightly between different design standards taking into account either the clear width between the flanges or webs, or the flat width considering the curved corner radii, as shown in Figure 3.7 and Figure 4.3.

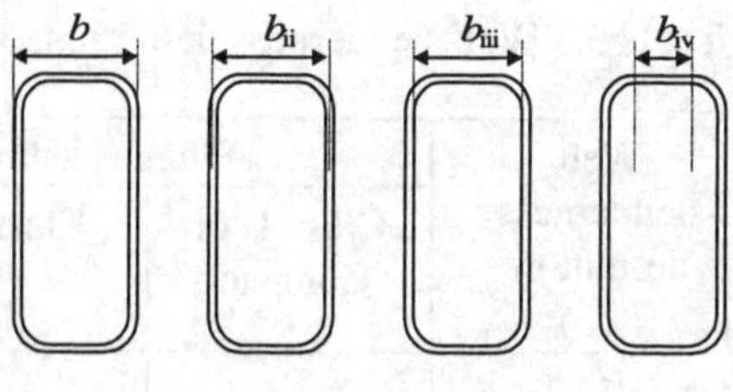

Figure 3.7 Different element width definitions for RHS flanges.

For example, AS 4100 considers the clear width $b$-$2t$ or $d$–$2t$, while AISC uses the flat width of the flange or web, $b$-$2r_{ext}$ or $d$-$2r_{ext}$. Tables 3.2, 3.3 and 3.4 show the variety of slenderness limits for the flange, width and diameter of hollow sections.

Table 3.2 Flange slenderness definitions

| Specification | Flange slenderness definition ($\lambda_f$) | Flange slenderness limits | | |
|---|---|---|---|---|
| | | Class 1, or Compact | Class 2 | Class 3, or Non-Compact |
| AS 4100 | $\frac{b-2t}{t}\sqrt{\frac{f_y}{250}}$ | 30 | - | 40 |
| Eurocode 3 | $\frac{b-2r_{ext}}{t}$ | $33\varepsilon'$ | $38\varepsilon'$ | $42\varepsilon'$ |
| AISC LRFD HSS | $\frac{b-2r_{ext}}{t}$ | $0.939\sqrt{\frac{E}{f_y}}$ | - | $1.40\sqrt{\frac{E}{f_y}}$ |
| BS 5950 | $\frac{b-5t}{t}$ | $26\varepsilon$ but $\leq$ $72\varepsilon-(d\text{-}5t)/t$ | $28\varepsilon$ but $\leq$ $54\varepsilon-0.5(d\text{-}5t)/t$ | $35\varepsilon$ |

*Notes*: (1) $\varepsilon = \sqrt{(275/p_y)}$ for use in BS 5950.
(2) $\varepsilon' = \sqrt{(235/p_y)}$ for use in Eurocode 3.
(3) The BS 5950 term ($b$-$5t$) is equivalent to ($b$-$2r_{ext}$) assuming $r_{ext} = 2.5t$.

Table 3.3 Web slenderness definitions

| Specification | Web slenderness definition ($\lambda_w$) | Flange slenderness limits | | |
|---|---|---|---|---|
| | | Class 1, or Compact | Class 2 | Class 3, or Non-Compact |
| AS 4100 | $\frac{d-2t}{t}\sqrt{\frac{f_y}{250}}$ | 82 | - | 115 |
| Eurocode 3 | $\frac{d-2r_{ext}}{t}$ | $72\varepsilon'$ | $83\varepsilon'$ | $124\varepsilon'$ |
| AISC LRFD HSS | $\frac{d-2r_{ext}}{t}$ | $3.76\sqrt{\frac{E}{f_y}}$ | - | $5.70\sqrt{\frac{E}{f_y}}$ |
| BS 5950 | $\frac{d-5t}{t}$ | $56\varepsilon$ | $70\varepsilon$ | $105\varepsilon$ |

*Notes*: (1) $\varepsilon = \sqrt{(275/p_y)}$ for use in BS 5950.
(2) $\varepsilon' = \sqrt{(235/p_y)}$ for use in Eurocode 3.
(3) The BS 5950 term $(b-5t)$ is equivalent to $(b-2r_{ext})$ assuming $r_{ext} = 2.5t$.

Table 3.4 CHS slenderness definitions

| Specification | CHS slenderness definition ($\lambda_w$) | Slenderness limits for CHS | | |
|---|---|---|---|---|
| | | Class 1, or Compact | Class 2 | Class 3, or Non-Compact |
| AS 4100 | $\frac{d_o}{t}\frac{f_y}{250}$ | 50 | - | 120 |
| Eurocode 3 | $\frac{d_o}{t}$ | $50\varepsilon'^2$ | $70\varepsilon'^2$ | $90\varepsilon'^2$ |
| AISC LRFD HSS | $\frac{d_o}{t}$ | $0.0714\frac{E}{f_y}$ | - | $0.309\frac{E}{f_y}$ |
| BS 5950 | $\frac{d_o}{t}$ | $40\varepsilon^2$ | $50\varepsilon^2$ | $140\varepsilon^2$ |

*Notes*: (1) $\varepsilon = \sqrt{(275/p_y)}$ for use in BS 5950.
(2) $\varepsilon' = \sqrt{(235/p_y)}$ for use in Eurocode 3.

### *3.2.5 Design Rules for Strength*

Once the classification of the section has been determined, the cross-sectional strength, or section capacity can be established.

#### *3.2.5.1 AS 4100*

AS 4100 determines the nominal section capacity as

$$M_s = f_y Z_e$$

where the calculation of the effective section modulus, $Z_e$, depends on the classification of the section. The element slenderness $\lambda_e$ of both the flange and the web are compared with the yield and plastic slenderness limits $\lambda_{ey}$ and $\lambda_{ep}$, to determine the classification of the section. The section slenderness $\lambda_s$ is determined from the element (flange or web) is more critical, having the higher value of $\lambda_e/\lambda_{ey}$.

The design section capacity is then determined by using the capacity factor $\phi = 0.9$ to give $\phi M_s$.

For compact sections, Clause 5.2.3 defines

$$Z_e = \min[S, 1.5Z] \tag{3.5}$$

where $Z$ and $S$ are the elastic and plastic section moduli, respectively.

For non-compact sections, Clause 5.2.4 gives:

$$Z_e = Z + \left(\frac{\lambda_{sy} - \lambda_s}{\lambda_{sy} - \lambda_{sp}}\right)(Z_c - Z) \tag{3.6}$$

where

$Z$ = elastic section modulus

$Z_c$ = $\min[S, 1.5Z]$

For slender sections square and rectangular hollow sections, Clause 5.2.5 defines

$$Z_e = Z\left(\frac{\lambda_{sy}}{\lambda_s}\right) \tag{3.7}$$

or for the effective cross-section determined by omitting from each flat compression element the width in excess of the width corresponding to $\lambda_{sy}$.

For slender circular hollow sections, Clause 5.2.5 defines

$$Z_e = \min\left[Z\sqrt{\frac{\lambda_{sy}}{\lambda_s}}; Z\left(\frac{2\lambda_{sy}}{\lambda_s}\right)^2\right] \tag{3.8}$$

It can be shown that for cold formed CHS that the term $\sqrt{\lambda_{sy}/\lambda_s}$ is always less than $(2\lambda_{sy}/\lambda_s)^2$, since the second limit is for slender sections mainly fabricated from thin plates by welding.

*3.2.5.2 BS 5950*

For all Class 1 and Class 2 Sections, Clause 4.2.5.2 defines

$$M_c = M_p = f_y S \tag{3.9}$$

where

$S$ = plastic section modulus

For all Class 3 sections, Clauses 3.5.6.3 and 4.2.5.2 give

$$M_c = f_y S_{eff} \text{ [or as a conservative alternative } M_c = f_y Z\text{]} \tag{3.10}$$

where

$Z$ = elastic section modulus

$S_{eff}$ = effective plastic section modulus

$$= \min\left[Z+(S-Z)\left(\frac{\dfrac{\beta_{3w}}{(d-5t)/t}-1}{\dfrac{\beta_{3w}}{\beta_{2w}}-1}\right), Z+(S-Z)\left(\frac{\dfrac{\beta_{3f}}{(d-5t)/t}-1}{\dfrac{\beta_{3f}}{\beta_{2f}}-1}\right)\right] \text{ for RHS}$$

$\beta_{2f}$, $\beta_{3f}$ = Class 2 & Class 3 limits for RHS flange

$\beta_{2w}$, $\beta_{3w}$ = Class 2 & Class 3 limits for RHS web

$$S_{eff} = Z + 1.485\left[\left(\frac{140}{d_o/t}\right)\left(\frac{275}{f_y}\right)^{0.5} - 1\right](S-Z) \text{ for CHS}$$

For all Class 4 sections, Clauses 3.6.2, 3.6.6 and 4.2.5.2 require the determination of effective widths to give

$$M_c = f_y Z_{eff} \tag{3.11}$$

where

$Z_{eff}$ = effective section modulus

= properties from the sections shown in Figure 3.8 for RHS

$$= Z\left(\frac{140}{d_o/t}\right)\left(\frac{275}{f_y}\right)^{0.25} \text{ for CHS}$$

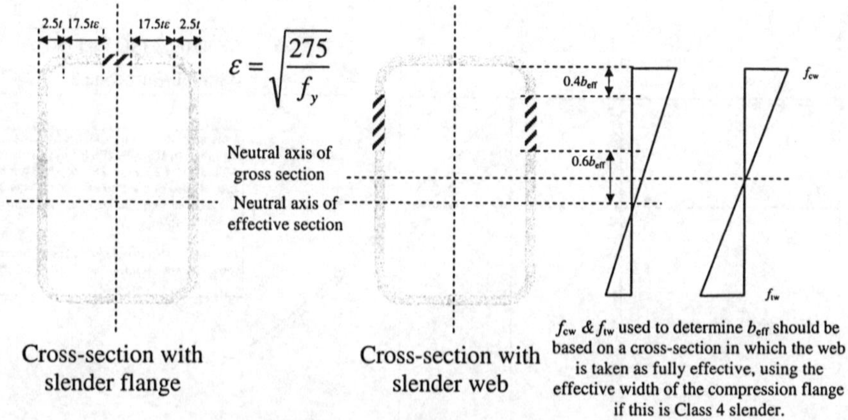


Figure 3.8 Definition of effective section properties for BS 5950

The effective width of the web compressive zone is

$$b_e = \frac{120\varepsilon t}{\left(1+\frac{f_{cw}-f_{tw}}{f_{yw}}\right)\left(1+\frac{f_{tw}}{f_{cw}}\right)} \tag{3.12}$$

where $f_{cw}$ is the maximum compressive stress in the web, $f_{tw}$ is the maximum tensile stress in the web and $f_{yw}$ is the yield stress of the web.

### *3.2.6 Comparison of Specifications*

The most obvious difference between specifications is the number of Classes. Eurocode 3 and BS 5950 have 4 classes, while AS 4100 and AISC LRFD have only 3. The AISC LRFD and AS 4100 class of "Non-Compact" incorporates the Eurocode 3 and BS 5950 Class 2 and Class 3.

Figure 3.9 below shows a simplified comparison of the section moment capacities calculated to the different specifications. In AS 4100 and AISC LFRD the moment capacity for non-compact sections degrade linearly between the plastic and yield moments. Hence AS 4100 and AISC LFRD exhibits conservatism for Class 2 sections which can reach the plastic moment with limited rotation capacity. Eurocode 3 and BS 5950 are conservative for some Class 3 sections, as there is a sudden discontinuity of capacity between the yield and plastic moments when the sections moves from Class 2 to Class 3. BS 5950 does allow the option for some type of degradation between $M_p$ and $M_y$ for Class 3 sections, as does Eurocode 3 for the case of an RHS with a Class 3 web and a Class 1 or 2 flange. For Class 4 sections, the different formulations are complicated, and a direct general comparison is not possible – it is sufficient to say all 4 specifications have similar formulation based on effective width formulae. The current BS 5950 approach appears the best formulation as it maintains $M_p$ for Class 2 sections then reduces gradually to $M_y$ for Class 3.

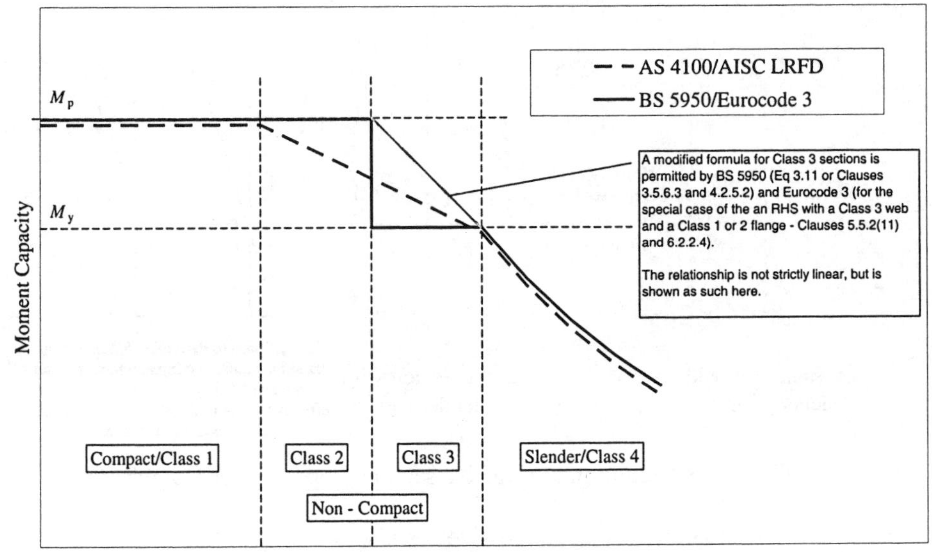


Figure 3.9 Comparison of section moment capacities

The Class 1 and Class 2 limits for flanges in BS 5950 in Table 3.2 also include an interaction term with the web slenderness. All other specifications use independent and separate limits for the flanges and webs. The test results of Wilkinson and Hancock (1998) suggested that interaction between the flange and web be considered when classifying cross sections in bending.

It is difficult to compare directly the slenderness limits in Tables 3.2, 3.3 and 3.4, since the different standards use different slenderness definitions. Hence Tables 3.5, 3.6 and 3.7 slightly modify the slenderness definition for each specification in order to give a meaningful comparison.

Table 3.5 Comparison of flange slenderness limits

| 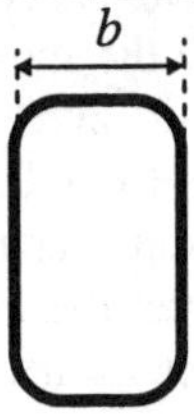  | Specification | Flange slenderness limits $b/t$ For full width $b$ and yield stress 350 N/mm$^2$ Assumes $r_{ext} = 2.5t$, $E = 200000$ N/mm$^2$ | | |
|---|---|---|---|---|
| | | Class 1, or Compact | Class 2 | Class 3, or Non-Compact |
| | AS 4100 | 27.4 | - | 35.8 |
| | Eurocode 3 | 32.0 | 36.1 | 39.4 |
| | AISC LRFD HSS | 27.4 | - | 38.5 |
| | BS 5950 | 28.0* | 29.8* | 36.0 |

Note: The BS 5950 limit does not include the interaction term considered in Table 3.1.

Table 3.6 Comparison of RHS web slenderness limits

| 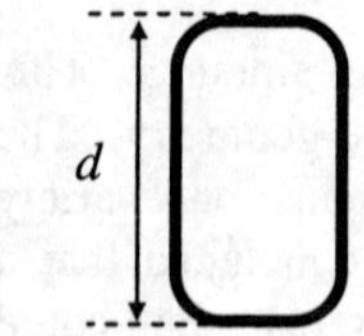 | Specification | Flange slenderness limits $d/t$ For full depth $d$ and yield stess 350 N/mm$^2$ Assumes $r_{ext} = 2.5t$, $E = 200000$ N/mm$^2$ | | |
|---|---|---|---|---|
| | | Class 1, or Compact | Class 2 | Class 3, or Non-Compact |
| | AS 4100 | 71.3 | - | 99.2 |
| | Eurocode 3 | 64.0 | 73.0 | 107 |
| | AISC LRFD HSS | 94.9 | - | 141 |
| | BS 5950 | 54.6 | 67.0 | 98.1 |

Table 3.7 Comparison of CHS slenderness limits

| 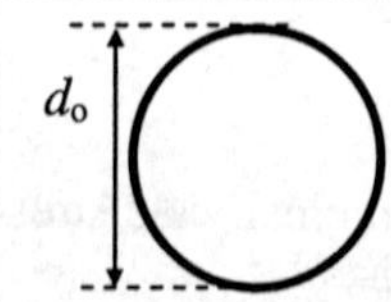 | Specification | CHS slenderness limits $d_o/t$ For yield stress 350 N/mm$^2$ Assumes $E = 200000$ N/mm$^2$ | | |
|---|---|---|---|---|
| | | Class 1, or Compact | Class 2 | Class 3, or Non-Compact |
| | AS 4100 | 35.7 | - | 85.7 |
| | Eurocode 3 | 33.6 | 47.0 | 60.4 |
| | AISC LRFD HSS | 40.8 | - | 177 |
| | BS 5950 | 31.4 | 39.2 | 110 |

The AISC LRFD limits are higher than the Australian and European counterparts, but it should be noted that AISC LRFD have a separate seismic specification (AISC 2002), in which stricter slenderness limits are applied. Some practising engineers may tend to use the lower seismic limits rather than the published compact limits when performing plastic design.

The BS 5950 limits are somewhat lower than the alternatives. BS 5950 specifies different limits for hot-formed hollow sections which are notable higher than those given in this Chapter which apply to cold-formed sections only. The BS 5950 slenderness limits for cold-formed tubular sections are generally 14 % more severe than the corresponding limits for hot formed tubular sections. AS 4100 also distinguishes between cold-formed and hot-rolled sections in its slenderness limits, with the cold-formed limits ranging from 0 % to 12 % lower than the equivalent hot-rolled limits. However the AS 4100 hot-rolled limits could be considered more applicable to traditional hot-rolled open sections rather than hot-formed tubes, since hot-formed tubes are not common in Australia.

AS 4100 is the only specification that does not permit plastic design of Compact/Class 1 cold-formed hollow sections, even though its compact limits are

based on sustaining the plastic moment for a suitably large rotation capacity. Plastic design is restricted to compact doubly-symmetric I-sections in AS 4100.

Plastic design requires that plastic hinges be able to rotate for certain amounts. This amount varies according the cross-section, the loading and the frame geometry. The Eurocode 3 Editorial Group (1989) summarise the rotation requirements for a variety of frames and multi-span beams constructed from I-sections and concluded that a value of $R = 3$ was a suitable value to ensure that a plastic collapse mechanism could form. The value of $R = 3$ is assumed in Eurocode 3 for Class 1 sections. The AISC LRFD specification is based on $R = 3$. AISC LRFD mentions [Commentary to Clause B.5] that in seismic regions greater rotation capacity may be required that the rotation provided by a compact section, and rotation capacities of the order R = 7 ~ 9 need to be provided. While AS 4100 does not permit plastic design for hollow sections, Hasan and Hancock (1988) and Zhao and Hancock (1991) used the value of $R = 4$ when recommending a compact limit for hollow section flanges. BS 5950 does not make specific reference to a rotation capacity requirement.

### *3.2.7 Examples*

Determine the section moment capacity of a cold-formed C350 (nominal yield stress of 350 N/mm$^2$) 100 × 50 × 2.5 RHS for bending about the major axis.

**Solution according to AS 4100**

1. Dimension and Properties

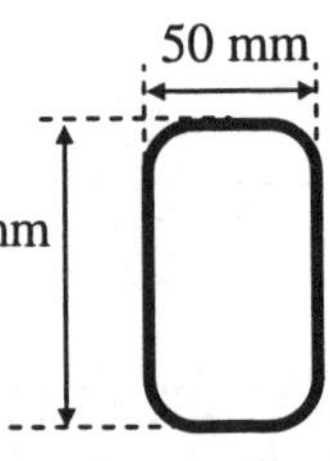


$b = 50$ mm
$d = 100$ mm
$t = 2.5$ mm
$A_n = 709$ mm$^2$
$Z_x = 18.2 \times 10^3$ mm$^3$
$S_x = 22.7 \times 10^3$ mm$^3$
$f_y = 350$ N/mm$^2$
$\phi = 0.9$

2. Classification

From Table 3.2, flange slenderness $\lambda_{ef} = \dfrac{b-2t}{t}\sqrt{\dfrac{f_y}{250}} = \dfrac{50-2\times2.5}{2.5}\sqrt{\dfrac{350}{250}} = 21.29$

Flange slenderness limits, $\lambda_{efp} = 30$ (plastic); $\lambda_{efy} = 40$ (yield)
Hence the flange is compact.

From Table 3.3, web slenderness $\lambda_{ef} = \dfrac{d-2t}{t}\sqrt{\dfrac{f_y}{250}} = \dfrac{100-2\times2.5}{2.5}\sqrt{\dfrac{350}{250}} = 44.93$

Web slenderness limits, $\lambda_{ewp} = 82$ (plastic); $\lambda_{efy} = 115$ (yield)
Hence the web is compact.

3. Moment Capacity

Using Equation (3.5)
Effective section modulus for a compact section, $Z_e = \min[S, 1.5Z] = 22.7 \times 10^3$ mm$^3$
Nominal moment capacity, $M_s = f_y Z_e = 22.7 \times 10^3 \times 350 = 7.95$ kNm
Design moment capacity, $\phi M_s = 0.9 \times 7.95 = 7.15$ kNm

**Solution according to BS 5950 Part 1**

1. Dimension and Properties

$b = 50$ mm
$d = 100$ mm
$t = 2.5$ mm
$A_n = 709$ mm$^2$
$Z_x = 18.2 \times 10^3$ mm$^3$
$S_x = 22.7 \times 10^3$ mm$^3$
$p_y = 350$ N/mm$^2$
$\varepsilon = (275/p_y)^{0.5} = 0.886$

2. Classification

From Table 3.2, flange slenderness
$(b\text{-}5t)/t = 37.5/2.5 = 15$

Class 1 limit
Either $\beta_{1f} = 26\varepsilon = 23.0$,
or $\beta_{1f} = 72\varepsilon - (d\text{-}5t)/t = 28.8$

Hence the flange is Class 1.

From Table 3.2, web slenderness
$(d\text{-}5t)/t = 87.5/2.5 = 35$
Class 1 limit, $\beta_{1f} = 56\varepsilon = 49.62$
Hence the web is Class 1.

The section is Class 1.

3. Moment Capacity

Using Equation (3.9), section capacity is
$M_c = p_y S = 350 \times 22.7 \times 10^3 = 7.95$ kNm

**Discussion**

This was a simple example and the calculations might be considered trivial. However the most commonly produced RHS/SHS are Class 1/Compact for major axis bending

as they represent efficient material use in bending. Examination of the range from one Australian manufacturer revealed the following:

- Grade C350 CHS: 57 sizes, 67 % compact, 33 % non-compact, 0 % slender,
- Grade C450 SHS: 87 sizes, 65 % compact, 20 % non-compact, 15 % slender,
- Grade C450 RHS: 80 sizes, 80 % compact, 17 % non-compact, 3 % slender, for $x$ axis bending; 33 % compact, 33 % non-compact, 33 % slender, for $y$ axis bending.

It is uncommon for slender sections to be used in bending applications, hence designers should not be concerned by the complexity of the slender design rules, as they are rarely needed in economical design situations.

## 3.3 Flexural-Torsional Buckling and Member Capacity

### *3.3.1 Flexural-Torsional Buckling*

When a beam is being bent about its major axis, flexural torsional buckling may occur. As shown in Figure 3.10, the beam deflects downwards, but at some stage buckling occurs over the length of the member, in which the cross section moves laterally (out of the plane of bending) and twists. The buckling deformations create bending about the minor axis and occur over the entire length of the beam, and hence this is sometimes called a member buckle, and the associated strength is sometimes called a member strength. Flexural-torsional buckling is also called lateral buckling, lateral-torsional buckling, or out-of-plane buckling. Figure 3.11 shows the flexural-torsional buckling of an RHS under experimental conditions.

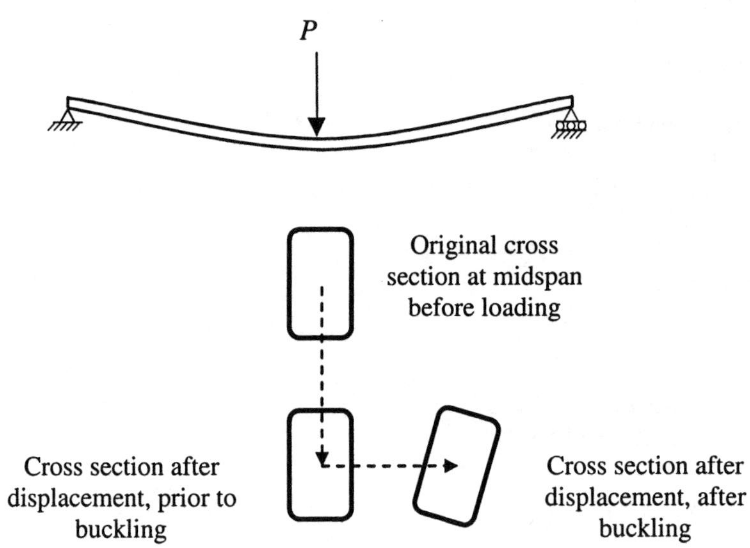


Figure 3.10 Buckling deflections of an RHS after flexural torsional buckling

Figure 3.11 Flexural torsional buckling of an RHS beam

Using an energy analogy, lateral buckling occurs because the strain energy associated with the twisting and out-of-plane bending deformations associated with the buckling deformations is less than the released potential energy of the applied load. RHS bending about the minor principal axis do not experience flexural torsional buckling, and only under extreme conditions do the symmetric CHS or SHS experience lateral buckling. Hence it is not considered an ultimate limit state for those sections in design specifications.

### *3.3.2 Critical Elastic Buckling Moment and Buckling of Real Beams*

It can be shown for the perfectly elastic and straight simply supported RHS beam in Figure 3.12 that experiences uniform bending moment over the entire length that the elastic flexural-torsional buckling moment is:

$$M_o = \sqrt{\left(\frac{\pi^2 EI_y}{L^2}\right)\left(GJ + \left(\frac{\pi^2 EI_w}{L^2}\right)\right)} = \sqrt{\frac{\pi^2 EI_y GJ}{L^2}} \text{ for RHS, since } I_w \approx 0 \text{ for hollow sections} \quad (3.13)$$

Further, $GJ$ is usually high for RHS, and hence $M_o$ is normally high. As the aspect ratio $d/b$ increases, $M_o$ usually reduces and lateral buckling may occur for sections of this type.

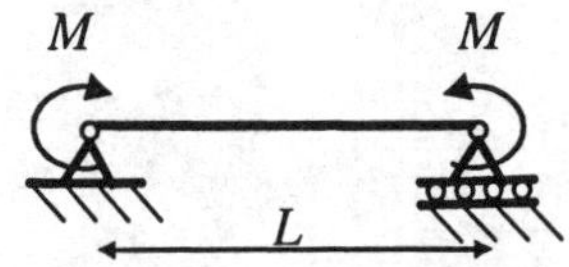


Figure 3.12 Simply supported beam with uniform bending moment

Several modifications are required to obtain the strength of real beams from the idealised behaviour of perfect beams.

It is clear that the buckling moment decreases with length. For short beams, the strength will be limited to a maximum value of the section capacity, ie it is limited by yielding and/or local buckling. Other factors, such as material non-linearity, residual stress and initial imperfections of the beam will further affect the capacity of the beam. This is illustrated in Figure 3.13 which shows the strength of a beam as a function of its length, with respect to the original section capacity. This is similar to the column curves which are used to determine the member capacity in compression, highlighted in Chapter 4.

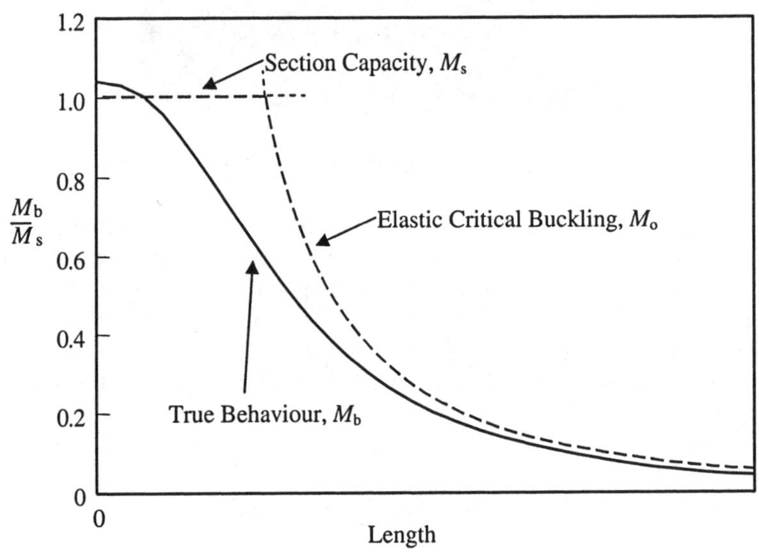


Figure 3.13 Comparison of elastic buckling, section capacity and behaviour of real beam with increasing length

The specimen in Figure 3.11 showed that the buckled shape of the beam involved lateral deflection, lateral rotation (about the vertical axis) and twisting (about the longitudinal axis). Different connections at the ends of beams (such a simple pinned shear connections compared to fully welded end plate moment connections) will

provide different levels of restraint against these three types of deformation. Intermediate connections from transverse beams, purlins, fly bracing etc must also be considered. In nearly all situations the restraint of the compression side of the beam is significant (sometimes it is referred to the critical flange). A detailed description of the effect of supports and the type of restraints they provide against lateral buckling can be found in Trahair *et al* (1993).

If the gravity load acts on the top of the beam, it can induce additional twisting as shown in Figure 3.14.

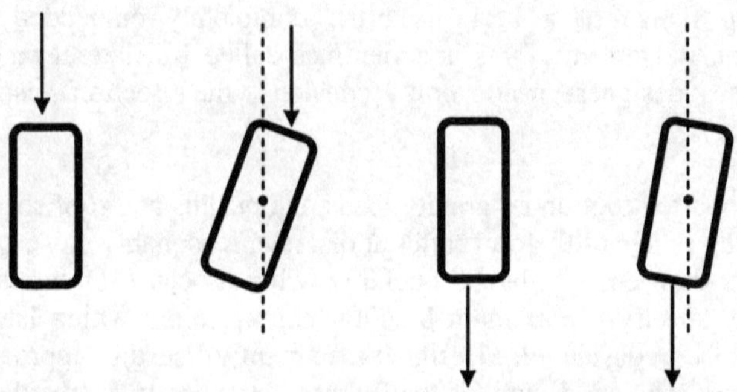


Figure 3.14 Top flange and shear centre loading

The combination of factors relating to the restraint of the beam, and the load height, are usually considered by calculating an effective length of the beam, determined from the original length of the beam multiplied by various correction factors.

The shape of the bending moment diagram can significantly affect the lateral buckling of a beam. The elastic buckling moment was derived for a beam experiencing uniform bending moment which is the most critical condition for lateral buckling. Figure 3.15 shows three beams each with the same magnitude of maximum bending moment in the beam. However the lateral buckling capacities of the beams will vary considerably, since only a small part of the beam experiences the peak moment. For the parabolic bending moment distribution, the buckling moment can be up to about 30 % higher than the uniform moment case. For the case of double curvature (the sign of the bending moment changes), the buckling moment could be almost 2.5 times higher than the uniform case. Double curvature beams have relatively high lateral buckling resistances since the change from compression to tensile stress in the flange resists deformation of that flange.

The usual approach in design specifications is to calculate the bending capacity assuming uniform moment, and then use a multiplier which is a function of the bending moment shape.

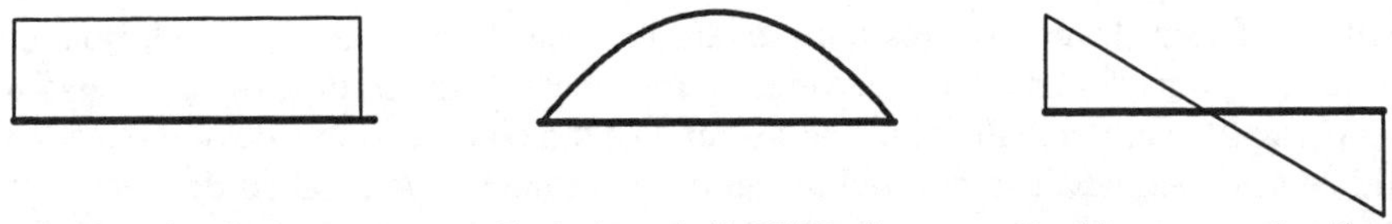


Figure 3.15 Variety of bending moment distributions

In some situations, lateral buckling will not occur. In the case of a composite flooring system, the top flange of a beam is often completely embedded in concrete, preventing lateral movement. This is sometimes called full lateral restraint, and in such cases the designer need only consider the section capacity (local buckling/yielding).

For a simply supported roof under gravity loads, the purlins and roof sheeting provide restraint to the top flange of the roof rafter at discrete, reasonably close spaces. If this bracing length is short enough, lateral buckling will not occur. Design specifications often commonly specify a maximum bracing length, below which lateral buckling does not need to be considered. Designers frequently use this approach, as it can provide for a more efficient design (as the full section capacity can be developed) and reduces calculation time.

### *3.3.3 Research Basis for Flexural Torsional Buckling*

The majority of research related to lateral buckling refers to investigations of open sections – and mainly I-sections. A comprehensive summary is available in Trahair (1993).

Initial lateral buckling design of hollow sections was performed in accordance with I-section rules. However, hollow sections are considerably different to I-sections with respect to the properties affecting lateral buckling. Hollow sections do not warp significantly, and the torsion constant, $J$, of an RHS is an order of magnitude higher than that of a comparable I-section. The manufacturing process of cold-formed hollow sections usually produces considerably smaller overall initial member crookedness imperfections.

Even though some specifications, such as BS 5950, use multiple beam curves for different sections, there was concern that it was still producing significantly conservative estimates of strength. Experimental investigations such as Zhao et al (1995a) and analytical and finite element investigations (Pi and Trahair 1995, Zhao et al 1995b) indicated that much higher strengths could be permitted. Some standards, such as AS 4100, have not yet included these recommendations, while others have. As an example, AISC LRFD does not even consider lateral buckling, as the Commentary to Clause 5.1 states

> "Because of the high torsional resistance of the closed cross-section, the critical unbraced lengths ... that correspond to the development of the plastic moment and the yield moment, respectively, are very large. For example ... an HSS 20×4×5/16 [508×102×8 RHS], which has one of the

largest depth-width ratios among standard HSS, has $L_p$ [length below which lateral buckling is assumed to not occur] of 8.7 ft [2.65 m] and $L_r$ [length at which inelastic lateral buckling changes to elastic lateral buckling] of 137 ft [41.7 m] as determined in accordance with the LRFD Specification. An extreme deflection limit might correspond to a length-to-depth ratio of 24 or a length of 40 ft [12.2 m] for this member. Using the specified linear reduction between the plastic moment and the yield moment for lateral-torsional buckling, the plastic moment is reduced by only 7 percent for the 40-ft length. In most practical designs where the moment gradient is also a factor, the reduction will be nonexistent or insignificant."

### *3.3.4 Design Rules for Member Strength*

It is assumed that the section capacity in bending has already been determined and that the designer has determined the restraint conditions of the beam, and the bending moment distribution of the beam segments between restraints. It is beyond the scope of this chapter to consider the type of restraints afforded by different connections – Trahair *et al* (1993) provides a substantial source of information.

#### *3.3.4.1 AS 4100*

Maximum bracing length for sections with full lateral restraint

Clause 5.3.2.4 allows for segments with full or partial restraints at both ends, where the length of the segment satisfies

$$l/r_y < (1800+1500\,\beta_m)(b/d)(250/f_y) \quad \text{(for RHS)} \qquad (3.14)$$

to be assumed to have full lateral restraint, and hence the section capacity may be used. $\beta_m$ may be taken as

- $\beta_m$ = -1.0 (conservative option)
- $\beta_m$ = -0.8 (for segments with transverse loads)
- $\beta_m$ = the ratio of the smaller to the larger end moment in the segment if there are no transverse loads. The value is positive for segments in reverse (double) curvature and negative for segments bent in single curvature.

Design of members without full lateral restraint

The effective length of the member in bending should be calculated first by considering the nature of the restraints (Clause 5.4), the load height, and then applying Clause 5.6.3, which gives:

$$l_e = k_t\, k_l\, k_r\, l \qquad (3.15)$$

where

$k_t$ = twist restraint factor

$k_l$ = load height factor

$k_r$ = lateral rotation restraint factor
$l$ = segment length

The elastic buckling moment is then determined using Equation 5.6.1.1(3) of AS 4100

$$M_o = \sqrt{\frac{\pi^2 EI_y GJ}{l^2}} \qquad (3.16)$$

The slenderness reduction factor ($\alpha_s$) is calculated from Equation 5.6.1.1(2) of AS 4100

$$\alpha_s = 0.6\left\{\sqrt{\left(\frac{M_s}{M_o}\right)^2 + 3} - \left(\frac{M_s}{M_o}\right)\right\} \qquad (3.17)$$

The effect of the bending moment distribution is covered by Clause 5.6.1.1(a) of AS 4100 which gives three options for determining $\alpha_m$, the moment modification factor:

1. $\alpha_m = 1.0$ (This is unnecessarily conservative in most cases)
2. Matching the bending moment shape to one of those in Tables 5.6.1 or 5.6.2 of AS 4100. This is often the easiest method.
3. $\alpha_m = \dfrac{1.7M^*_m}{\sqrt{(M^*_2)^2 + (M^*_3)^2 + (M^*_4)^2}}$, where $M^*_m$ is the maximum bending moment in the segment, and $M^*_2$, $M^*_3$, $M^*_4$ are the values of bending moment at the quarter points of the segment. This method is commonly used in computerized design methods.

The member moment capacity is then given by

$$M_{bx} = \min[M_{sx}, \alpha_m \alpha_s M_{sx}] \qquad (3.18)$$

The design member moment capacity $\phi M_{bx}$ is obtained using the capacity factor $\phi = 0.9$.

*3.3.4.2 BS 5950*

The approach of BS 5950 is the same as AS 4100, though the exact formulation of the steps is naturally different.

Maximum bracing length for sections with full lateral restraint

Clause 4.3.6.1 allows for beams where the effective length term $L_E/r_y$ is less than some constant multiplied by $(275/p_y)$ to be designed for the section capacity only (local buckling). This limit is a function of the aspect ratio of the cross-section and varies from $770 \times (275/p_y)$ for $d/b = 1.25$ to $170 \times (275/p_y)$ for $d/b = 4.0$.

Design of members without full lateral restraint

The effective length of the member ($L_E$) in bending should be calculated first using the factors in Clause 4.3.5 which take into account the nature of the restraints.

The equivalent slenderness $\lambda_{LT}$ is determined from Clause B.2.6.1 and is related to the elastic critical buckling moment.

$$\lambda_{LT} = 2.25(\phi_b \lambda \beta_w)^{0.5}$$

where

$$\phi_b = \left(\frac{S_x^2 \gamma_b}{AJ}\right)^{0.5}$$

$$\gamma_b = \left(1 - \frac{I_y}{I_x}\right)\left(1 - \frac{J}{2.6 I_x}\right) \tag{3.19}$$

$$\lambda = \frac{L_E}{r_y}$$

$$\beta_w = \frac{M_c}{M_p} = \left(\frac{\text{section capacity}}{\text{plastic moment}}\right)$$

The bending strength ($p_b$) is then determined from the yield stress ($p_y$) and the modified slenderness ($\lambda_{LT}$) by either using a table such as Table 17 of BS 5950, or by the method highlighted in Clause B2.1.

The buckling resistance moment ($M_b$) is calculated using the same equations as the section capacity equations given above, but using $p_b$ rather than $p_y$.

$M_b = p_b S_x$ for Class 1 and 2 sections

$M_b = p_b S_{x,eff}$ or $p_b Z_x$ for Class 3 sections (3.20)

$M_b = p_b Z_{x,eff}$ or Class 4 sections

The ratio $p_b/p_y$ could be considered equivalent to the slenderness reduction factor $\alpha_s$ in AS 4100.

The effect of the bending moment distribution is covered by Clause 4.3.6.6 of BS 5950, in which the equivalent uniform moment factor $m_{LT}$ is determined by matching the bending moment shape to one of those in Table 18 or

$$m_{LT} = \max\left[0.2 + \frac{M_2 + M_3 + M_4}{M_{max}}, 0.44\right] \tag{3.21}$$

where $M_{max}$ is the maximum bending moment in the segment, and $M_2$, $M_3$, $M_4$ are the values of bending moment at the quarter points of the segment. $m_{LT}$ is the equivalent of the reciprocal of the AS 4100 term $\alpha_m$.

The member moment capacity is then taken as the lesser of the $M_c$ and $M_b/m_{LT}$.

### *3.3.5 Comparison of Specifications*

AISC LRFD takes a most practical approach in ignoring lateral buckling for RHS. As was outlined above, extraordinarily long beam lengths are required to make the deductions for lateral buckling significant.

AS 4100 and BS 5950 require that lateral buckling be considered in bending design, but it will often be found (as in the case of the next example), that the lateral buckling capacity is the same or almost the same as the original section capacity (local buckling).

Lateral buckling design in AS 4100 is identical to the design of I-sections, since there is only one beam design curve. It is recognised that cold-formed hollow sections generally are straighter than I-sections, hence a higher beam curve could be justified. BS 5950 has two beam curves, one for "rolled" sections and a lower one for "welded" sections (accounting for the generally higher residual stresses and imperfections). Since the straighter cold-formed sections are not treated separately, this may produce slightly conservative answers. Eurocode 3 contains 4 recommended beam design curves, and rolled I-sections form the highest two curves. Cold-formed RHS are not specifically covered, and it is left to the designer to select an appropriate curve, but since there is no recommendation, most designers would probably pick the unnecessarily conservative assumption of the lowest beam curve.

### *3.3.6 Examples*

Consider a 100 × 50 × 2.5 C350 RHS which forms a simply supported beam of length 5.3 m from which some roofing and associated services are being suspended. The member only has full restraint at the ends, but no lateral rotation restraints. What is the maximum uniformly distributed load that the member can resist?

**Solution according to AS 4100**

Draw the simply supported beam and the bending moment diagram.

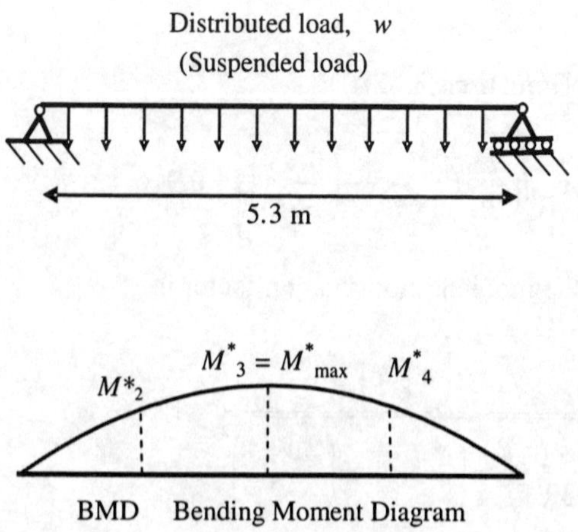


$$M^*_2 = M^*_4 = \left(\frac{wL}{2}\right)\frac{L}{4} - \left(\frac{wL}{4}\right)\frac{L}{8} = \frac{3wL^2}{32}$$

$$M^*_m = M^*_3 = \frac{wL^2}{8}$$

Fig. 3.16 Bending moment distribution in simply supported beam

1. The relevant properties of the 100 × 50 × 2.5 C350 RHS are as follows.

$b = 50\,\text{mm}$ $\quad d = 100\,\text{mm}$ $\quad t = 2.5\,\text{mm}$

$f_y = 350\,\text{MPa}$ $\quad E = 200000\,\text{MPa}$ $\quad G = 80000\,\text{MPa}$

$Z_{ex} = 22.7\times10^3\,\text{mm}^3$ $\quad M_{sx} = f_y Z_{ex} = 7.95\,\text{kNm}$

$\phi = 0.9$ $\quad \phi M_{sx} = 7.15\,\text{kNm}$

$I_y = 0.311\times10^6\,\text{mm}^4$ $\quad J = 0.754\times10^6\,\text{mm}^4$ $\quad r_y = 20.9\,\text{mm}$

2. The maximum bracing length (Equation (3.14))

$l_{max} = r_y(1800+1500\beta_m)(b/d)(250/f_y) = 20.9\times(1800+1500\times(-0.8))(0.5)(250/350)$
$= 4.47\,\text{m}$

Since the beam is 5.3 m long, it does not satisfy the maximum bracing length requirement, and hence design for lateral buckling needs to be considered.

3. Compute effective length $l_e$ (Equation (3.15))

The segment under consideration is the full beam with length $l$.

Twist restraint factor (AS 4100 Table 5.6.3(1)): $k_t = 1.0$
(full restraint at ends)

Load height factor (AS 4100 Table 5.6.3(2)): $k_\ell = 1.0$
(load at or below centroid)

Lateral rotation restraint factor (AS 4100 Table 5.6.3(3)): $k_r = 1.0$
(no lateral rotation restraint)

$l_e = k_t k_\ell k_r l = 5300\,\text{mm}$

4. Elastic buckling moment (Equation (3.16))

$$M_o = \sqrt{\left(\frac{\pi^2 EI_y}{l_e^2}\right)GJ} = \sqrt{\left(\frac{\pi^2\times200000\times0.311\times10^6}{5300^2}\right)\times80000\times0.754\times10^6} = 36.3\,\text{kNm}$$

5. The slenderness reduction factor is (Equation (3.17))

$$\alpha_s = 0.6\left[\sqrt{\left(\frac{M_{sx}}{M_o}\right)^2 + 3} - \left(\frac{M_{sx}}{M_o}\right)\right] = 0.6\left[\sqrt{\left(\frac{7.95}{36.3}\right)^2 + 3} - \left(\frac{7.95}{36.3}\right)\right] = 0.916$$

6. Referring to the bending moment, the moment modification factor is

$$\alpha_m = \frac{1.7M_m^*}{\sqrt{(M_2^*)^2 + (M_3^*)^2 + (M_4^*)^2}} = \frac{1.7\left(\frac{wL^2}{8}\right)}{\sqrt{\left(\frac{3wL^2}{32}\right)^2 + \left(\frac{wL^2}{8}\right)^2 + \left(\frac{3wL^2}{32}\right)^2}} = 1.166$$

7. The member capacity is (Equation (3.18))

$$M_{bx} = \min[\alpha_m \alpha_s M_{sx}, M_{sx}] = \min[1.166 \times 0.916 \times 7.95, 7.95] = 7.95 \text{ kNm}$$

$$\phi M_{bx} = 0.9 \times 7.95 = 7.15 \text{ kNm}$$

There is no reduction for the possibility of lateral buckling in this case (even though the beam did not originally satisfy the maximum bracing length requirements). For design purposes, $\phi M_{bx} > M^* = w^* L^2/8$.

Hence, the maximum uniformly distributed factored load is given by

$$w^* = \frac{8\phi M_{bx}}{L^2} = \frac{8 \times 7.15}{5.5^2} = 1.90 \text{ kN/m}$$

**Solution according to BS 5950**

The properties and bending moment diagram were calculated previously.

The maximum bracing length
$L_{E,\max} = 340 r_y \times (275/p_y) = 340 \times 20.9 \times 275/350 = 5.58$ m

Table 13 of BS 5950 gives for this case of the simply supported beam, with an unrestrained compression flange and normal loading (not destabilizing) that the effective length should be $L_E = L + 2d = 5.5$ m in this case, which is less than the maximum length. Hence design for lateral buckling does not need to be considered.

Hence the section capacity may be used and the same load is in the case of AS 4100 can be resisted.

$$M_{cx} = 7.95 \text{ kNm}$$

**Discussion**

In this example, the length of the beam was of the order of $300r_y$. The length of 5.3 m was deliberately chosen to illustrate the different methods between AS 4100 and BS 5950. There was a slight (10 %) reduction in strength to account for lateral buckling according to AS 4100 (which was then counteracted by an increase in capacity due to the shape of the bending moment). Even a beam twice as long, $600r_y$, would only experience a strength decrease of 22 % (not including the possible benefits of bending moment shape). For a similarly dimensioned I-section, there is a 75 % strength reduction due to lateral buckling when the beam length is $300r_y$, and an 85 % reduction for $600r_y$.

The load of 1.92 kN/m would in fact induce an elastic midspan deflection of approximately 100 mm, which is equivalent to *l*/50 which is a very large deflection for a beam. This further illustrates that in most practical bending situations, lateral buckling of RHS is not a significant factor to consider.

A proposed design curve was given in Zhao et al (1995b) for RHS beams with uniform moment ($\alpha_m = 1.0$):

$$M_{bx} = (1.056 - 0.278\lambda^2) \cdot M_{sx} \quad \text{for } 0.45 \le \lambda \le 1.40$$

$$M_{bx} = M_o \quad \text{for } \lambda > 1.40$$

where

$$\lambda = \sqrt{\frac{M_{sx}}{M_o}}$$

$$M_o = \frac{\pi}{l}\sqrt{EI_y GJ}$$

The beam length corresponding to $\lambda$ of 0.45 is the maximum bracing length ($l_{max}$).

Setting $\lambda_p = 0.45$, $l_{max}$ can be written as:

$$l_{max} = \frac{\pi\sqrt{EI_y GJ}}{M_{sx}} \cdot \lambda_p^2$$

For the Example 3.3.6,
$l_{max} = 4.9$ m

It should be mentioned that the proposed formula in Zhao et al (1995b) was based on the results for 75 × 25 × 2.5 RHS which has a *b*/*d* ratio of 1/3 whereas the section in this example, 100 × 50 × 2.5 RHS, has *b*/*d* of 0.5.

## 3.4 References


1. AISC (1999), Load and Resistance Factor Design Specification for Structural Steel Buildings, (AISC LRFD), American Institute of Steel Construction, Chicago, Illinois, USA
2. AISC (2000), Load and Resistance Factor Design Specification for Steel Hollow Structural Sections, (AISC LRFD), American Institute of Steel Construction, Chicago, Illinois, USA
3. AISC (2002), Seismic Provisions for Structural Steel Buildings, (AISC LRFD), American Institute of Steel Construction, Chicago, Illinois, USA
4. Bleich, F. (1952), *Buckling Strength of Metal Structures*, Engineering Societies Monographs, McGraw-Hill, New York, USA
5. BSI (2000), Structural use of Steelwork in Building, BS 5950, Part 1, British Standard Institution, London, UK
6. Canadian Standards Association (2001), CAN/CSA-S16-01: Limits States Design of Steel Structures, Toronto, Ontario, Canada
7. Corona, E. and Vaze, S.P. (1996), Buckling of Elastic Plastic Square Tubes Under Bending, *International Journal of Mechanical Sciences*, Elsevier, Vol 38, No 7, pp 753 - 775
8. Dawe, J.L., and Kulak, G.L. (1984a), Plate Instability of W Shapes, *Journal of Structural Engineering*, American Society of Civil Engineers, Vol 110, No 6, June 1984, pp 1278-1291
9. Dawe, J.L., and Kulak, G.L. (1984b), Local Buckling of W Shape Columns and Beams, *Journal of Structural Engineering*, American Society of Civil Engineers, Vol 110, No 6, June 1984, pp 1292-134
10. Dawe, J.L., and Kulak, G.L. (1986), Local Buckling Behaviour of Beam-Columns, *Journal of Structural Engineering*, American Society of Civil Engineers, Vol 112, No 11, November 1986, pp 2447-2461
11. Elchalakani, M., Zhao, X.L. and Grzebieta, R.H. (2002a), Plastic Slenderness Limit for Cold-Formed Circular Steel Hollow Sections, *Australian Journal of Structural Engineering*, Vol. 3, No. 3, pp. 127-139
12. Elchalakani, M., Zhao, X.L. and Grzebieta, R.H. (2002b), Bending Tests to Determine Slenderness Limits for Cold-Formed Circular Hollow Sections, *Journal of Constructional Steel Research*, Vol. 58, No. 11, pp. 1407-1430
13. Eurocode 3 Editorial Group (1989), The *b/t* Ratios Controlling the Applicability of Analysis Models in Eurocode 3, *Document 5.02*, Background Documentation to Chapter 5 of Eurocode 3, Aachen University, Germany
14. EC3 (2003), Eurocode 3: Design of Steel Structures, Part 1-1: General Rules and Rules for Buildings, prEN 1993-1-1:2003, November 2003, European Committee for Standardisation, Brussels, Belgium
15. Galambos, T.V. (1968), *Structural Members and Frames*, Prentice-Hall Series in Structural Analysis and Design (W. J. Hall, editor), Prentice-Hall, London, U.K
16. Galambos, T.V. (1976), Proposed Criteria for Load and Resistance Factor Design of Steel Building Structures, *Research Report No 45*, Civil Engineering Department, Washington University, St. Louis, Mo., USA. (Also published as American Iron and Steel Institute (AISI), *Bulletin No 27*, January 1978)
17. Haaijer, G. and Thurlimann, B. (1958), On Inelastic Buckling in Steel, *Journal of the Engineering Mechanics Division*, Proceedings of the American Society of Civil Engineers, Vol 84. No EM 2, April 1958, Proceedings Paper No 1581

18. Hasan, S.W., and Hancock, G.J. (1988), Plastic Bending Tests of Cold-Formed Rectangular Hollow Sections, *Research Report*, No R586, School of Civil and Mining Engineering, The University of Sydney, Sydney, Australia. (also published in *Steel Construction*, Journal of the Australian Institute of Steel Construction, Vol 23, No 4, November 1989, pp 2-19.
19. Johnston, B. (Ed.) (1976), *Guide to Stability Design Criteria for Metal Structures*, Structural Stability Research Council, 3rd edition, John Wiley
20. Jouaux, R. (2004), The plastic behaviour of cold-formed SHS (square hollow sections) under bending and compression, *Diploma Thesis*, Department of Civil Engineering, The University of Sydney / Fakultät für Bauingenieur-, Geo-, und Umweltwissenschaften, Universität Karlsruhe, 2004
21. Kato, B. (1965), Buckling Strength of Plates in the Plastic Range, *Publications*, International Association for Bridge and Structural Engineering, Vol 25, 1965, pp 127 - 141
22. Korol, R.M., and Hudoba, J. (1972), Plastic Behaviour of Hollow Structural Sections, *Journal of the Structural Division*, American Society of Civil Engineers, Vol 98, No ST5, pp 1007-1023
23. Lay, M.G. (1965), Flange Local Buckling in Wide-Flange Shapes, *Journal of the Structural Division*, American Society of Civil Engineers, Vol 91, No ST6, pp 95 - 116, December 1965
24. Lukey, A.F. and Adams, P.F. (1969), Rotation Capacity of Beams Under Moment Gradient, *Journal of the Structural Division*, American Society of Civil Engineers, Vol 95, No ST6, pp 1173 - 1188
25. Sherman, D.R. (1986), Inelastic Flexural Buckling of Cylinders, *Steel Structures, Recent Research Advance and Their Applications to Design*, Proceedings of the invited papers for the International Conference, Pavlovic, M. N. (ed), Budva, Yugoslavia, 29 September-1 October, pp 339-357
26. Ostapenko, A. (1983), Local Buckling, *Structural Steel Design*, Chapter 17, (L. Tall editor), 2nd edition, Robert Kreiger Publishing, Malabar, Florida, USA
27. Pi, Y.L. and Trahair, N.S. (1995), Lateral buckling strengths of cold-formed rectangular hollow sections, *Thin-Walled Structures*, Volume 22, Issue 2, pp 71-95
28. Standards Australia (1998), Australian Standard *AS 4100 Steel Structures*, Standards Australia, Sydney, Australia
29. Stranghöner, N. (1995), Untersuchungen zum Rotationsverhakten von Trägern aus Hohloprofilen, *PhD Thesis*, Institute of Steel Construction, University of Technology, Aachen, Germany
30. Timoshenko, S. and Gere, J. (1969), *Theory of Elastic Stability*, 2nd edition, McGraw-Hill, New York, New York, USA
31. Trahair N.S., Hogan T.J. and Syam, A.A. (1993), Design of Unbraced Beams, *Steel Construction*, Journal of the Australian Institute of Steel Construction, Vol 27, No 1, pp 2 - 26
32. Trahair N.S. (1993), *Flexural Torsional Buckling of Structures*, EF & N Spon, London
33. Ueda, Y. and Tall, L. (1967), Inelastic Buckling of Plates with Residual Stresses, *Publications*, International Association for Bridge and Structural Engineering, Vol 27, 1967, pp 211 - 254
34. Wilkinson T. and Hancock G.J. (1998), Tests to examine the compact web slenderness of cold-formed RHS, *Journal of Structural Engineering*, American Society of Civil Engineers, Vol 124, No 10, October 1998, pp 1166-1174

35. Wilkinson T. (2003), Recommendations for Cold-Formed RHS in Bending and Compression, in: *Tubular Structures X*, Proceedings of the 10th International Symposium on Tubular Structures, Madrid Spain, Jaurrieta, Alonso & Chica eds., Balkema, Rotterdam, The Netherlands, pp 293-300
36. Zhao, X.L. & Hancock, G.J. (1991), Tests to Determine Plate Slenderness Limits for Cold-Formed Rectangular Hollow Sections of Grade C450, *Steel Construction*, Journal of Australian Institute of Steel Construction, Vol 25, No 4, November 1991, pp 2-16
37. Zhao, X.L., Hancock, G.J., and Trahair, N.S. (1995a), Lateral Buckling Tests of Cold-Formed RHS Beams, *Journal of Structural Engineering*, American Society of Civil Engineers, Vol 121, No 11, November 1995, pp 1565-1573
38. Zhao, X.L., Hancock, G.J., Trahair, N.S. and Pi, Y.L. (1995b), Lateral Buckling of RHS Beams, In: *Structural Stability and Design*, Kitipornchai, S., Hancock, G.J. and Bradford, M. (eds), Balkema: Rotterdam, The Netherlands, pp 55-60

# Chapter 4: Members Subjected to Compression

## 4.1 General

Compression members are commonly used as columns in building structures, chords or webs in trusses, bridge piers or braces in framed structures. The maximum strength of a steel compression member depends, to a large extent, on the member length and end support conditions. Steel compression members are commonly classified as short, intermediate or long members. Each range has its own characteristic type of behaviour. For example, local buckling and/or yielding is the major concern for short compression members while overall buckling affects the strength of long members. Intermediate members are more complex to analyze but also are the most common in steel structures. They may fail by a combination of yielding, overall buckling and/or local buckling including interaction between buckling modes. A schematic relationship between the maximum strength of a compression member and its length is shown in Figure 4.1. Section 4.2 deals with short members where local buckling or material yielding is the key issue. Section 4.3 deals with intermediate members where the interaction of local buckling and overall buckling may occur. The behaviour of long members is not described here since it has been well documented as Euler buckling in the literatures, and can be found in standard steel design texts (e.g. Trahair and Bradford 1988).

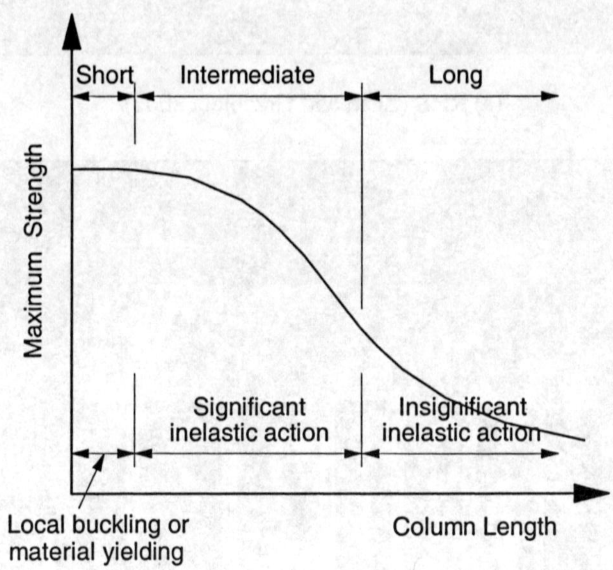


Figure 4.1 A schematic view of strength versus length

## 4.2 Section Capacity

### *4.2.1 Local Buckling in Compression*

Local buckling occurs if the width-to-thickness ratio or diameter-to-thickness ratio of a tube exceeds a certain value (see Section 4.2.2). Typical inelastic local buckling modes are shown in Figure 4.2. Those in (a) are often called a "roof mechanism" and those in (b) an "elephant's foot" buckle.

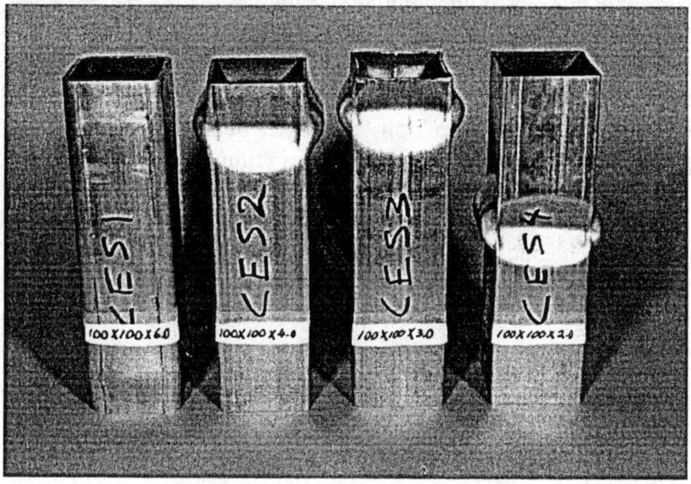


(a) RHS (Zhao and Grzebieta 2002)

(b) CHS (Zhao et al 2002)

Figure 4.2 Typical local buckling modes of cold-formed tubes

Local buckling may occur before yielding of the whole cross-section thus preventing the section from reaching its full axial capacity. The effect of local buckling on section capacity is considered in design standards using different approaches. For example, the concept of form factor ($k_f$) is used in AS 4100 (Standards Australia 1998). The form factor is defined as the ratio of effective area of a locally buckled section to gross area. The effective area is calculated using the effective widths for the flat elements of RHS or the effective outside diameter for CHS. A similar concept, called the $Q$ factor, is used in the AISC specification (AISC 2000) to take into account the effect of local buckling. The effective area of the section based on the effective width concept is used directly in BS 5950 Part 1 (BSI 2000) and Eurocode 3 (2003) without introducing another reduction factor as in AS 4100 and the AISC specification.

### 4.2.2 *Limiting Width-to-Thickness Ratios*

The concept of plate element slenderness ($\lambda_e$) is used in AS 4100 whereas the concept of width-to-thickness ratio or diameter-to-thickness ratio is used in the AISC LRFD Specification, BS 5950 Part 1 and Eurocode 3 for the purpose of design. The relationship between the plate element slenderness and the width or diameter-to-thickness ratio can be generally expressed as:

$$\lambda_e = (\frac{\text{width}}{\text{thickness}}) \cdot \sqrt{\frac{f_y}{250}} \text{ for RHS} \tag{4.1}$$

$$\lambda_e = (\frac{\text{diameter}}{\text{thickness}}) \cdot (\frac{f_y}{250}) \text{ for CHS} \tag{4.2}$$

The plate element slenderness takes account of the yield stress so that a higher yield stress produces a plate element slenderness which is higher and can be compared with a fixed value. By comparison, the AISC Specification, BS 5950 and Eurocode 3 compare the actual width/thickness ratio with a limit which varies inversely as the square root of the yield stress. Both methods result in the same outcome when determining slender sections. Clear width ($b$-$2t$) is used in AS 4100 whereas the flat width ($b$-$2r_{ext}$) is used in AISC specification, BS 5950 and Eurocode 3. The symbols $b$, $t$ and $r_{ext}$ are defined in Figure 2.2. The clear width and the flat width are defined in Figure 4.3.

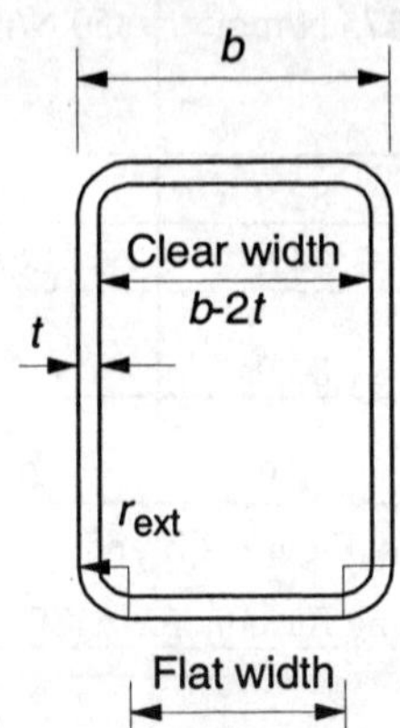


Figure 4.3 Definition of clear width and flat width

Limiting values on $\lambda_e$ or width-to-thickness ratios are specified in different standards. They were based on research all over the world (SSRC 1991, Beedle 1992, Sherman 1992, Hancock and Zhao 1992, Rondal et al 1996). In order to show a clear comparison among standards, all the limits are converted to the limiting overall width-to-thickness ($b/t$) or diameter-to-thickness ($d_o/t$) ratios. The comparisons are given in Table 4.1 for RHS and Table 4.2 for CHS. Some examples are given in the tables for tubes with a yield stress of 275, 350, 355 and 450 N/mm$^2$. The limiting values from various standards are very close, i.e. within 3% for RHS and within 10% for CHS.

Table 4.1 Limiting width-to-thickness ratios for cold-formed RHS

| Design Standard | Expression of limiting overall width-to-thickness ratio ($b/t$) | $b/t$ limit for $f_y$ of 275 N/mm$^2$ | $b/t$ limit for $f_y$ of 350 N/mm$^2$ | $b/t$ limit for $f_y$ of 355 N/mm$^2$ | $b/t$ limit for $f_y$ of 450 N/mm$^2$ |
|---|---|---|---|---|---|
| AISC LRFD | $\frac{2 \cdot r_{ext}}{t} + 39.6 \cdot \varepsilon$ | 42.8 | 38.5 | 38.2 | 34.5 |
| AS 4100 | $2 + 40 \cdot \varepsilon$ | 40.1 | 35.8 | 35.6 | 31.8 |
| BS 5950 Part 1 | $5 + 36.7 \cdot \varepsilon$ | 40.0 | 36.0 | 35.8 | 32.4 |
| Eurocode 3 Part 1.1 | $\frac{2 \cdot r_{ext}}{t} + 40.7 \cdot \varepsilon$ | 43.8 | 39.4 | 39.2 | 35.3 |

Note: $b$ is the overall width defined in Figure 2.2, $\varepsilon = \sqrt{250/f_y}$ where $f_y$ is in N/mm$^2$, $r_{ext}$ is the external corner radius defined in Figure 2.2. A value of $2.5t$ is used for $r_{ext}$ in the calculation. $E$ = 200,000 N/mm$^2$ is used in deriving formula for AISC (1999).

Table 4.2 Limiting diameter-to-thickness ratios for cold-formed CHS

| Design Standard | Expression of limiting diameter-to-thickness ratio ($d_o/t$) | $d_o/t$ limit for $f_y$ of 275 N/mm$^2$ | $d_o/t$ limit for $f_y$ of 350 N/mm$^2$ | $d_o/t$ limit for $f_y$ of 355 N/mm$^2$ | $d_o/t$ limit for $f_y$ of 450 N/mm$^2$ |
|---|---|---|---|---|---|
| AISC LRFD | $91.2 \cdot \varepsilon^2$ | 82.9 | 65.1 | 64.2 | 50.7 |
| AS 4100 | $82 \cdot \varepsilon^2$ | 74.5 | 58.6 | 57.8 | 45.6 |
| BS 5950 Part 1 | $88 \cdot \varepsilon^2$ | 80.0 | 62.9 | 62.0 | 48.9 |
| Eurocode 3 Part 1.1 | $84.6 \cdot \varepsilon^2$ | 76.9 | 60.4 | 59.6 | 47.0 |

Note: $d_o$ is the diameter defined in Figure 2.2 and $\varepsilon = \sqrt{250/f_y}$ where $f_y$ is in N/mm$^2$. $E$ = 200,000 N/mm$^2$ is used in deriving formula for AISC (1999).

### 4.2.3 *Design Section Capacity*

As mentioned in Chapter 1, design examples will be given in this book in accordance with AS 4100 and BS 5950 Part 1. Only design formulae in these two standards will be discussed further. The section dimensions defined in Figure 2.2 are used throughout the book. The symbols for capacity, area and yield stress are the same as those used in AS 4100 and BS 5950.

*4.2.3.1 AS 4100*

The design section capacity ($\phi N_s$) according to AS 4100 can be written as:

$$\phi \cdot N_s = \phi \cdot k_f \cdot A_n \cdot f_y \qquad (4.3)$$

where $\phi$ (=0.9) is the capacity factor, $N_s$ is the nominal section capacity, $k_f$ is the form factor, $A_n$ is the net area of the cross-section and $f_y$ is the yield stress.

The form factor for SHS, RHS and CHS can be derived as follows. It is assumed in the derivation that SHS and RHS have squared corners (see Figure 4.4) since the clear width ($b$-$2t$) is used in AS 4100.

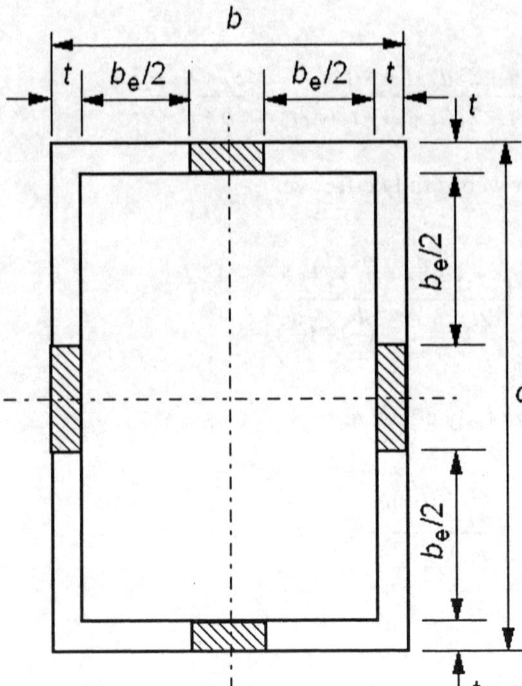


Figure 4.4 Approximated dimensions used to derive $k_f$ for a cold-formed RHS in AS 4100

**For SHS**

$$k_{f,SHS} = \frac{4 \cdot b_e \cdot t + 4 \cdot t^2}{4 \cdot (b-2t) \cdot t + 4 \cdot t^2} = \frac{b_e + t}{b - t}$$

where $b_e = (b-2t) \cdot \frac{\lambda_{ey}}{\lambda_e} = (b-2t) \cdot \frac{\lambda_{ey}}{(\frac{b-2t}{t}) \cdot \sqrt{\frac{f_y}{250}}} = \lambda_{ey} \cdot \sqrt{\frac{250}{f_y}} \cdot t = 40 \cdot \varepsilon \cdot t$

The value of 40 is given in Table 5.2 in AS 4100 as the plate element slenderness limit ($\lambda_{ey}$).

Hence

$$k_{f,SHS} = \frac{40 \cdot \varepsilon \cdot t + t}{b - t} = \frac{40 \cdot \varepsilon + 1}{\frac{b}{t} - 1} \approx \frac{40\varepsilon + 2}{\frac{b}{t}} = \frac{(\frac{b}{t})_{limit}}{(\frac{b}{t})} \tag{4.4}$$

where $(\frac{b}{t})_{limit}$ for AS 4100 is given for a range of yield stress values in Table 4.1.

**For RHS**

$$k_{f,RHS} = \frac{2 \cdot b_e \cdot t + 2 \cdot d_e \cdot t + 4 \cdot t^2}{2 \cdot (b-2t) \cdot t + 2 \cdot (d-2t) \cdot t + 4 \cdot t^2} = \frac{b_e + d_e + 2t}{b + d - 2t}$$

If neither flange nor web is fully effective,

$$k_{f,RHS} = \frac{40 \cdot \varepsilon \cdot t + 40 \cdot \varepsilon \cdot t + 2t}{b + d - 2t} \approx \frac{2 \cdot (\frac{b}{t})_{limit}}{(\frac{b}{t}) + (\frac{d}{t})} \tag{4.5}$$

If only the web is not fully effective,

$$k_{f,RHS} = \frac{b + 40 \cdot \varepsilon \cdot t}{b + d} = \frac{(\frac{b}{t}) + (\frac{b}{t})_{limit}}{(\frac{b}{t}) + (\frac{d}{t})} \tag{4.6}$$

**For CHS**

According to AS 4100, the effective outside diameter ($d_e$) for a CHS shall be the lesser of

$$d_e = d_o \cdot \sqrt{\frac{\lambda_{ey}}{\lambda_e}} \le d_o \quad \text{and}$$

$$d_e = d_o \cdot \left(\frac{3 \cdot \lambda_{ey}}{\lambda_e}\right)^2 \le d_o$$

It can be demonstrated that for normal cold-formed CHS $\sqrt{\frac{\lambda_{ey}}{\lambda_e}}$ is always less than $\left(\frac{3 \cdot \lambda_{ey}}{\lambda_e}\right)^2$ since the second limit is for slender sections mainly fabricated from thin-plates by welding.

$$k_{f,CHS} = \frac{\frac{\pi}{4} \cdot d_e^2 - \frac{\pi}{4} \cdot (d_e - 2t)^2}{\frac{\pi}{4} \cdot d_o^2 - \frac{\pi}{4} \cdot (d_o - 2t)^2} = \frac{d_e^2 - (d_e - 2t)^2}{d_o^2 - (d_o - 2t)^2} = \frac{d_e - t}{d_o - t} \approx \frac{d_e}{d_o}$$

where

$$d_e = d_o \cdot \sqrt{\frac{\lambda_{ey}}{\lambda_e}} = d_o \cdot \sqrt{\frac{\lambda_{ey}}{\frac{d_o}{t} \cdot \frac{f_y}{250}}} = d_o \cdot \sqrt{\frac{82 \cdot \varepsilon^2}{\frac{d_o}{t}}} = d_o \cdot \sqrt{\frac{(\frac{d_o}{t})_{limit}}{(\frac{d_o}{t})}}$$

The value of 82 is given in Table 5.2 in AS 4100 as the plate element slenderness limit ($\lambda_{ey}$).

Hence

$$k_{f,CHS} \approx \sqrt{\frac{(\frac{d_o}{t})_{limit}}{(\frac{d_o}{t})}} \tag{4.7}$$

where $(\frac{d_o}{t})_{limit}$ for AS 4100 is given for a range of yield stress values in Table 4.2.

*4.2.3.2 BS 5950 Part 1*

The compression resistance ($P_c$) of a cross-section according to BS 5950 Part 1 can be written as:

$$P_c = A_{eff} \cdot p_c \tag{4.8}$$

where $A_{eff}$ is the effective area and $p_c$ is the compressive strength.

The effective area for SHS, RHS and CHS can be derived as follows.

**For SHS and RHS**

An explicit expression for $A_{eff}$ can be derived using the effective width concept in BS 5950 Part 1, as shown in Figure 4.3 where the term ε is defined in Table 4.1 with $\varepsilon = \sqrt{250/f_y}$.

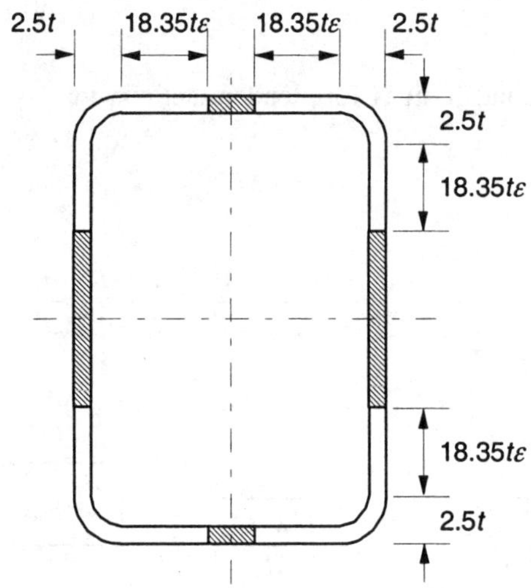


Figure 4.5 Dimensions used to derive $A_{eff}$ for a cold-formed RHS

If neither flange nor web is fully effective,

$$A_{eff} = 8\cdot[(18.35\cdot t\cdot\varepsilon)\cdot t] + \pi\cdot[(2.5t)^2 - (1.5t)^2] = 4\cdot(36.7\cdot\varepsilon)\cdot t^2 + 4\cdot\pi\cdot t^2$$

It can be simplified as

$$A_{eff} = 4\cdot t^2\cdot\left\{\left(\frac{b}{t}\right)_{limit} - 5 + \pi\right\} \tag{4.9}$$

where the $(\frac{b}{t})_{limit} = 5 + 36.7\cdot\varepsilon$ as given in Table 4.1 with $\varepsilon = \sqrt{250/f_y}$, which is converted from Table 12 in BS 5950 Part 1.

If only the web is not fully effective,

$$A_{eff} = 4\cdot[(18.35\cdot t\cdot\varepsilon)\cdot t] + 2\cdot(b-5t)\cdot t + \pi\cdot[(2.5t)^2 - (1.5t)^2]$$

It can be simplified as

$$A_{eff} = 4\cdot t^2\cdot\left\{0.5\cdot\left(\frac{b}{t}\right) + 0.5\cdot\left(\frac{b}{t}\right)_{limit} - 5 + \pi\right\} \tag{4.10}$$

**For CHS**

According to BS 5950Part 1, the effective area for CHS can be expressed as:

$$A_{eff} = \frac{\pi}{4}[d_o^2 - (d_o - 2t)^2] \cdot \sqrt{(\frac{80}{d_o / t}) \cdot (\frac{275}{f_y})}$$

which can be simplified as

$$A_{eff} = \frac{\pi}{4}[d_o^2 - (d_o - 2t)^2] \cdot \sqrt{\frac{(\frac{d_o}{t})_{limit}}{(\frac{d_o}{t})}} \tag{4.11}$$

where $(\frac{d_o}{t})_{limit} = 88 \cdot \varepsilon^2$ as given in Table 4.2 with $\varepsilon = \sqrt{250 / f_y}$ , which is converted from Table 12 in BS 5950 Part 1.

### 4.2.4 *Examples*

Determine the compressive section capacity of a cold-formed C350 (nominal yield stress of 350 N/mm$^2$) 100×50×2.5 RHS.

**Solution according to AS 4100**

1. Dimensions and Properties

$b$ = 50 mm
$d$ = 100 mm
$t$ = 2.5 mm
$A_n$ = 709 mm$^2$ based on actual rounded corners
$f_y$ = 350 N/mm$^2$
$\phi$ = 0.9

2. Form factor

$b/t = 50/2.5 = 20 < (b/t)_{limit}$ of 35.8 given in Table 4.1
Flange is fully effective

$d/t = 100/2.5 = 40 > (b/t)_{limit}$ of 35.8 given in Table 4.1
Web is not fully effective

From Equation (4.6)

$$k_{f,RHS} = \frac{(\frac{b}{t}) + (\frac{b}{t})_{limit}}{(\frac{b}{t}) + (\frac{d}{t})} = \frac{20 + 35.8}{20 + 40} = 0.93$$

3. Nominal section capacity

$N_s = k_f A_n f_y = 0.93 \times 709 \times 350 = 230{,}780$ N = 231 kN

4. Design section capacity

$\phi N_s = 0.9 \times 231 = 208$ kN

**Solution according to BS 5950 Part 1**

1. Dimension and Properties

$b = 50$ mm
$d = 100$ mm
$t = 2.5$ mm
$p_y = 350$ N/mm$^2$

2. Effective area

$b/t = 50/2.5 = 20 < (b/t)_{limit}$ of 36.0 given in Table 4.1
Flange is fully effective

$d/t = 100/2.5 = 40 > (b/t)_{limit}$ of 36.0 given in Table 4.1
Web is not fully effective

From Equation (4.10)

$$A_{eff} = 4 \cdot t^2 \cdot \left\{0.5 \cdot \left(\frac{b}{t}\right) + 0.5 \cdot \left(\frac{b}{t}\right)_{limit} - 5 + \pi\right\} = 4 \times 2.5^2 \cdot \{0.5 \times 20 + 0.5 \times 36 - 5 + \pi\}$$

$= 654$ mm$^2$

3. Compression strength

$P_c = A_{eff}\, p_c = A_{eff}\, p_y = 654 \times 350 = 228{,}900$ N = 229 kN

For the purpose of a simple comparison, the nominal capacity in AS 4100 of 231 kN can be compared with the compression strength of 229 kN in BS 5950 Part 1.

## 4.3 Member Capacity

### *4.3.1 Interaction of Local and Overall Buckling*

Steel columns can be classified as short, intermediate or long members. The maximum strength of short (stub) columns is equal to the section capacity as described in Section 4.2. For very long columns, the maximum strength is limited by the elastic overall (Euler) buckling capacity. The column capacity depends more on the bending stiffness of the member ($EI$), its length ($l$) and the restraint against rotation at the supports. The most commonly used columns in steel structures belong to those with an intermediate length. Their strength is affected by both local buckling of the cross-section and the overall (Euler) buckling of the member. It is often called the interaction of local and overall buckling. The maximum strength of such columns depends not only on the bending stiffness and length but also on the yield stress of the steel, the distribution of residual stress over the cross-section, the cross-section slenderness, and the magnitude of the initial imperfections in columns and component plates of the cross-section.

Typical failure modes of cold-formed tubular columns are shown in Figure 4.6 for an intermediate length column and Figure 4.7 for a long column respectively.

Figure 4.6 Typical failure mode of a cold-formed RHS column (Key 1988)

Figure 4.7 Typical failure mode of a cold-formed CHS column (Zhao et al 2000)

### 4.3.2 *Column Curves*

The concept of multiple column curves (Johnston 1976, Rotter 1982) has been adopted by most of the standards for column design. A column curve is normally plotted as non-dimensional axial capacity (i.e. a ratio of member capacity to section capacity, also called member slenderness reduction factor $\alpha_c$ in AS 4100) versus modified member slenderness $\lambda_n$. Different column curves correspond to different cross-section types, and distributions and magnitudes of the residual stresses. For example, two column curves are given in AS 4100 (Standards Australia 1998) for cold-formed RHS and CHS based on the research by Key et al (1988). They are plotted in Figure 4.8 with two different values of compression member section constant ($\alpha_b$). The upper curve ($\alpha_b$ = -1.0) is for cold-formed (stress relieved) RHS and CHS with form factor ($k_f$) of 1.0. The lower curve ($\alpha_b$ = -0.5) is for the following three cases: (i) cold-formed (non-stress relieved) RHS and CHS with form factor of 1.0, (ii) cold-formed (stress relieved) RHS and CHS with form factor less than 1.0, (iii) cold-formed (non-stress relieved) RHS and CHS with form factor less than 1.0. The same data are presented in a tabulated format in AS 4100.

Strut curve c is allocated for design of cold-formed RHS and CHS columns in BS 5950 Part 1. The data are presented in a tabulated format in BS 5950 Part 1 for different values of yield stress and also by formulae in Annex C. Examples are given in Figure 4.9 for yield stress values of 275 N/mm$^2$, 355 N/mm$^2$ and 460 N/mm$^2$.

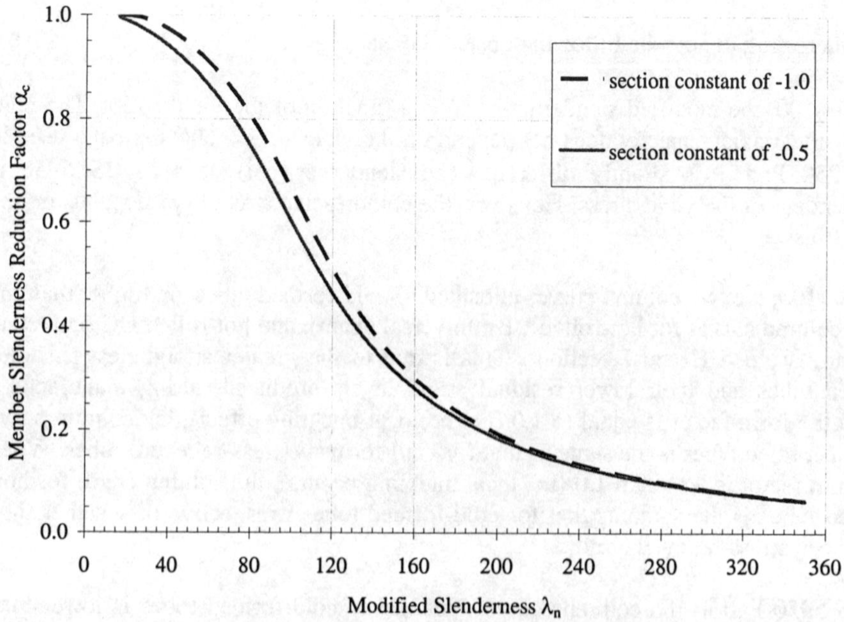


Figure 4.8 Column curves for cold-formed RHS and CHS given in AS 4100

(modified slenderness $\lambda_n = \sqrt{k_f} \cdot \frac{l_e}{r} \cdot \sqrt{\frac{f_y}{250}}$ versus

member slenderness reduction factor $\alpha_c = \frac{N_c}{N_s}$)

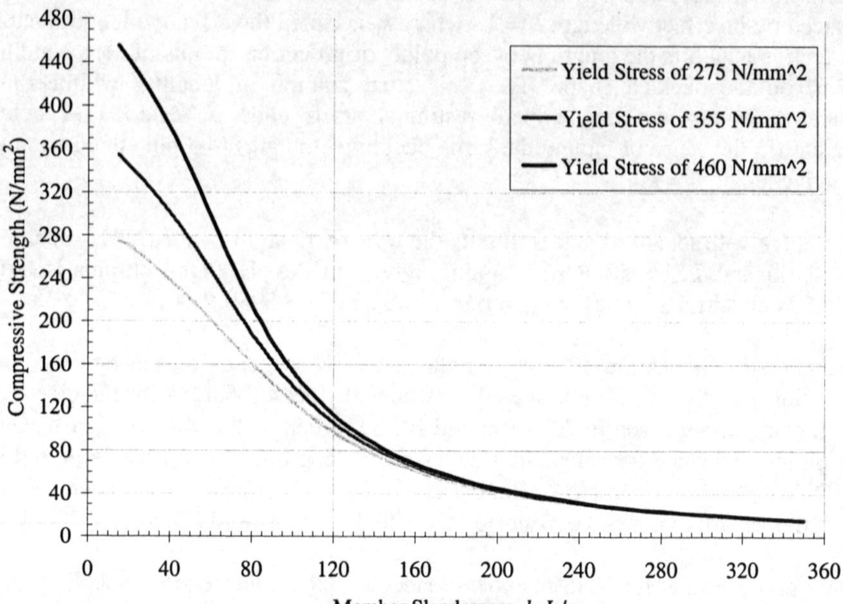


Figure 4.9 Column curves for cold-formed RHS and CHS given in BS 5950 Part 1 (yield stresses of 275 N/mm$^2$, 355 N/mm$^2$ and 460 N/mm$^2$)

It is interesting to note the following observations:

In AS 4100, the modified slenderness ($\lambda_n$) is a function of the yield stress. The non-dimensional axial capacity does not depend on the yield stress. The approach used in BS 5950 Part 1 is slightly different. The slenderness ($\lambda$) used in BS 5950 is independent of the yield stress. However, the compressive strength $p_c$ depends on the yield stress.

In AS 4100, the two column curves allocated to cold-formed tubes are higher than the three column curves for hot-rolled UB (universal beam), and hot-rolled UC (universal column), welded H and I sections. This is due to the greater straightness of cold-formed tubes and their lower residual stress levels produced during manufacture. When the form factor is equal to 1.0 (i.e. no local buckling effect), the column curve for hot-formed tubes is the same as that for cold-formed (stress-relieved) tubes. When the form factor is less than 1.0 (i.e. local buckling occurs), the column curve for hot-formed tubes is the same as that for cold-formed tubes irrespective of whether they have been stress-relieved or not.

In BS 5950 Part 1, the columns curve allocated to cold-formed tubes is lower than those allocated to hot-finished tubes, rolled I, H sections and welded sections.

### *4.3.3 Effective Length for Compression Members*

In order to use the design rules developed for buckling of a pin-ended column in situations where a column is restrained at its ends, the actual length of the column ($l$) is replaced by its effective length $l_e = k_e l$ where $k_e$ is called the effective length factor. The effective length is the length between points of inflection (points of zero bending moment) on the buckled shape. Thus an actual column of length $l$ with certain rotational restraints and translational restraints at its ends is regarded as being equivalent (in the sense of having the same buckling strength) to a pin ended column of effective length $l_e = k_e l$.

For members with idealised end restraints the effective length factor ($k_e$) is specified in AS 4100 and BS 5950 Part 1. Values given in AS 4100 are summarised in Table 4.3. Very similar values are given in Table 22 of BS 5950 Part 1.

For members in frames the effective length factor ($k_e$) depends on the ratios of the compression member stiffness to the end restraint stiffnesses. Charts for the effective length factor ($k_e$) are given in AS 4100 and BS 5950 Part 1. For members in welded lattice girders guidance for calculating the effective length factor ($k_e$) can be found in the CIDECT Design Guide No.2 (Rondal et al 1996). For members in space frames the effective length factor ($k_e$) depends on the nodal systems used (Zhao et al 2000).

It should be noted that for RHS members loaded in axial compression, buckling may take place about either principal axis. It is necessary to calculate the geometrical slenderness ratios $(l_e/r)_x$ and $(l_e/r)_x$ about the principal $x$ and $y$ axes respectively. The critical case is the one with a larger $(l_e/r)$ ratio where $r$ is the radius of gyration of an RHS.

Table 4.3 Effective length factors for members with idealised end restraints (from Table 4.6.3.2 of AS 4100)

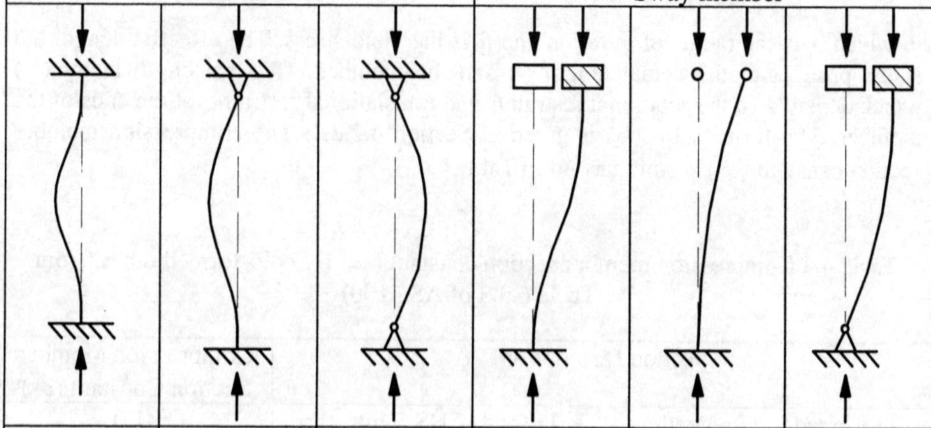

| Braced member | | | Sway member | | |
|---|---|---|---|---|---|
| $k_e = 0.7$ | $k_e = 0.85$ | $k_e = 0.1.0$ | $k_e = 1.2$ | $k_e = 2.2$ | $k_e = 2.2$ |

### 4.3.4 *Design Member Capacity*

*4.3.4.1 AS 4100*

The design member capacity ($\phi N_c$) according to AS 4100 can be written as:

$$\phi \cdot N_c = \phi \cdot \alpha_c \cdot N_s \tag{4.12}$$

where $\phi$ (=0.9) is the capacity factor, $N_s$ is the nominal section capacity defined in Section 4.2.3.1, $\alpha_c$ is the member slenderness reduction factor given in Figure 4.8 or by Equation (4.13).

$$\alpha_c = \xi \cdot \left\{ 1 - \sqrt{1 - \left( \frac{90}{\xi \cdot \lambda} \right)^2} \right\} \tag{4.13a}$$

$$\xi = \frac{\left( \frac{\lambda}{90} \right)^2 + 1 + \eta}{2 \cdot \left( \frac{\lambda}{90} \right)^2} \tag{4.13b}$$

$$\lambda = \lambda_n + \alpha_a \cdot \alpha_b \tag{4.13c}$$

$$\eta = 0.00326 \cdot (\lambda - 13.5) \geq 0 \tag{4.13d}$$

$$\lambda_n = \left( \frac{l_e}{r} \right) \cdot \sqrt{k_f} \cdot \sqrt{\frac{f_y}{250}} \tag{4.13e}$$

$$\alpha_a = \frac{2100 \cdot (\lambda_n - 13.5)}{\lambda_n^2 - 15.3 \cdot \lambda_n + 2050} \qquad (4.13f)$$

in which, $r$ is the radius of gyration and $f_y$ is the yield stress. The effective length ($l_e$) is the product of the actual length ($l$) and the member effective length factor ($k_e$) which depends on the rotational restraints and translational restraints at the ends of the member. The form factor ($k_f$) is given in Section 4.2.3.1. The compression member section constant ($\alpha_b$) is summarised in Table 4.4.

Table 4.4 Compression member section constant ($\alpha_b$) for cold-formed tubes (from Table 6.3.3 of AS 4100)

| Section Description | Compression Member Section Constant ($\alpha_b$) |
|---|---|
| cold-formed (stress relieved) RHS and CHS with form factor ($k_f$) of 1.0 | -1.0 |
| cold-formed (non-stress relieved) RHS and CHS with form factor of 1.0 | -0.5 |
| cold-formed (stress relieved) RHS and CHS with form factor less than 1.0 | -0.5 |
| cold-formed (non-stress relieved) RHS and CHS with form factor less than 1.0 | -0.5 |

*4.3.4.2 BS 5950 Part 1*

The compression resistance ($P_c$) of a member according to BS 5950 Part 1 can be written as:

For class 1 plastic, class 2 compact or class 3 semi-compact cross-sections:

$$P_c = A_g \cdot p_c \qquad (4.14a)$$

For class 4 slender cross-sections:

$$P_c = A_{eff} \cdot p_{cs} \qquad (4.14b)$$

where $A_g$ is the gross cross-sectional area, $A_{eff}$ is the effective cross-sectional area defined in Section 4.2.3.2, $p_c$ is the compressive strength given in a tabulated format in BS 5950 Part 1 (see Figure 4.9 for examples) or by Equation (4.15) and $p_{cs}$ is the value of $p_c$ for a reduced slenderness $\lambda \cdot \sqrt{\frac{A_{eff}}{A_g}}$ in which $\lambda = \frac{L_E}{r}$ where $L_E$ is the effective length. Note that the term $\sqrt{\frac{A_{eff}}{A_g}}$ is equivalent to $\sqrt{k_f}$ in AS 4100.

$$p_c = \frac{p_E \cdot f_y}{\phi + \sqrt{\phi^2 - p_E \cdot f_y}} \tag{4.15a}$$

$$\phi = \frac{f_y + (\eta + 1) \cdot p_E}{2} \tag{4.15b}$$

$$p_E = \frac{\pi^2 \cdot E}{\lambda^2} \tag{4.15c}$$

$$\lambda = \frac{k_e \cdot L}{r} \tag{4.15d}$$

$$\eta = \alpha \cdot (\lambda - \lambda_o)/1000 \text{ but } \eta \geq 0 \tag{4.15e}$$

$$\lambda_o = 0.2 \cdot \pi \cdot \sqrt{\frac{E}{f_y}} \tag{4.15f}$$

$\alpha = 5.5$ for strut curve (c), i.e. for design of cold-formed RHS and CHS.

### 4.3.5 *Examples*

#### *4.3.5.1 Example 1*

Determine the compressive member capacity of a cold-formed (non-stress relieved) 100×50×2.5 RHS column with a pin-ended length of 5300 mm. Assume that the yield stress is 350 N/mm$^2$.
If some bracing are provided to reduce the minor axis effective length to 2650 mm, determine the compressive member capacity.

**Solution according to AS 4100**

1. Dimension and Properties

$b = 100$ mm
$d = 50$ mm
$t = 2.5$ mm
$A_n = 709$ mm$^2$
$l = 5300$ mm
$r_x = 35.9$ mm
$r_y = 20.9$ mm
$f_y = 350$ N/mm$^2$
$\phi = 0.9$

2. Form factor

From the example in Section 4.2.4: $k_f = 0.93$

3. Nominal section capacity

From the example in Section 4.2.4: $N_s = 231$ kN

4. Modified member slenderness

$k_e = 1.0$ because of pin-ended condition

$$\lambda_{nx} = \left(\frac{l_{ex}}{r_x}\right) \cdot \sqrt{k_f} \cdot \sqrt{\frac{f_y}{250}} = \frac{1.0 \cdot 5300}{35.9} \cdot \sqrt{0.93} \cdot \sqrt{\frac{350}{250}} \approx 169$$

$$\lambda_{ny} = \left(\frac{l_{ey}}{r_y}\right) \cdot \sqrt{k_f} \cdot \sqrt{\frac{f_y}{250}} = \frac{1.0 \cdot 5300}{20.9} \cdot \sqrt{0.93} \cdot \sqrt{\frac{350}{250}} \approx 289$$

5. Member slenderness reduction factor

From Table 4.4, section constant $\alpha_b = -0.5$

From Figure 4.8, $\alpha_{cx} \approx 0.26$
From Figure 4.8, $\alpha_{cy} \approx 0.09$

6. Design member capacity

$\phi N_{cx} = 0.9 \times 0.26 \times 231 = 0.9 \times 60.06 \approx 54.0$ kN
$\phi N_{cy} = 0.9 \times 0.09 \times 231 = 0.9 \times 20.79 \approx 18.7$ kN

Therefore the compressive member capacity is 18.7 kN buckling about the minor ($y$) axis.

If some bracing are provided to reduce the minor axis effective length to 2650 mm, the modified member slenderness becomes

$$\lambda_{ny} = \left(\frac{l_{ey}}{r_y}\right) \cdot \sqrt{k_f} \cdot \sqrt{\frac{f_y}{250}} = \frac{2650}{20.9} \cdot \sqrt{0.93} \cdot \sqrt{\frac{350}{250}} \approx 145$$

From Figure 4.8, $\alpha_{cy} \approx 0.34$

$\phi N_{cy} = 0.9 \times 0.34 \times 231 \approx 70.7$ kN

Therefore the compressive member capacity is 54 kN buckling about the major ($x$) axis.

**Solution according to BS 5950 Part 1**

1. Dimensions and Properties

$b = 100$ mm
$d = 50$ mm
$t = 2.5$ mm
$A_g = 709$ mm$^2$
$L = 5300$ mm
$r_x = 35.9$ mm
$r_y = 20.9$ mm
$p_y = 350$ N/mm$^2$

2. Cross-sectional area

From the example in Section 3.2.7, the RHS 100×50×2.5 is Class 1 section. Hence the gross cross-sectional area $A_g$ should be used, i.e.

$A_g = 709$ mm$^2$

3. Member slenderness

$k_e = 1.0$ because of pin-ended condition

$$\lambda_x = \frac{L_{Ex}}{r_x} = \frac{1.0 \cdot 5300}{35.9} = 148$$

$$\lambda_y = \frac{L_{Ey}}{r_y} = \frac{1.0 \cdot 5300}{20.9} = 254$$

4. Compressive strength

It can be seen from Figure 4.9 that the influence of the yield stress on the column curves is not significant when the member slenderness is larger than 120. Therefore the curve for $p_y$ of 355 N/mm$^2$ is used in this example for tubes with a yield stress of 350 N/mm$^2$.

$p_{cx} \approx 75$ N/mm$^2$
$p_{cy} \approx 29$ N/mm$^2$

4. Member compression resistance

$P_{cx} = A_g\, p_{cx} = 709 \times 75 \approx 53{,}200$ N $= 53.2$ kN
$P_{cy} = A_g\, p_{cy} = 709 \times 29 \approx 20{,}500$ N $= 20.5$ kN

Therefore the compressive member capacity is 20.5 kN buckling about the minor ($y$) axis.

If some bracing are provided to reduce the minor axis effective length to 2650 mm, the member slenderness becomes

$$\lambda_y = \frac{L_{Ey}}{r_y} = \frac{2650}{20.9} = 127$$

Using the curve for $p_y$ of 355 N/mm$^2$ in Figure 4.9 as an approximation:

$$p_{cy} \approx 97 \ \text{N/mm}^2$$

$$P_{cy} = A_g\, p_{cy} = 709 \times 97 \approx 68{,}700 \text{ N} = 68.7 \text{ kN}$$

Therefore the compressive member capacity is 53.2 kN buckling about the major ($x$) axis.

*4.3.5.2 Example 2*

Determine the compressive member capacity of a cold-formed (non-stress relieved) 200×200×5 SHS column with a pin-ended length of 3600 mm. Assume that the yield stress is 355 N/mm$^2$.

**Solution according to AS 4100**

1. Dimensions and Properties

$b = 200$ mm
$d = 200$ mm
$t = 5.0$ mm
$A_n = 3810$ mm$^2$
$l = 3600$ mm
$r = 79$ mm
$f_y = 355$ N/mm$^2$
$\phi = 0.9$

2. Form factor

$b/t = 200/5 = 40 > (b/t)_{limit}$ of 35.6 given in Table 4.1
Flange is not fully effective

$d/t = 200/5 = 40 > (b/t)_{limit}$ of 35.6 given in Table 4.1
Web is not fully effective

From Equation (4.5),

$$k_{f,RHS} \approx \frac{2 \cdot (\frac{b}{t})_{limit}}{(\frac{b}{t}) + (\frac{d}{t})} = \frac{2 \cdot 35.6}{40 + 40} = 0.89$$

3. Nominal section capacity

$N_s = k_f A_n f_y = 0.89 \times 3810 \times 355 = 1{,}203{,}770$ N = 1204 kN

4. Modified member slenderness

$k_e$ = 1.0 because of pin-ended condition

$$\lambda_n = \left(\frac{l_e}{r}\right)\cdot\sqrt{k_f}\cdot\sqrt{\frac{f_y}{250}} = \frac{1.0\cdot 3600}{79}\cdot\sqrt{0.89}\cdot\sqrt{\frac{355}{250}} \approx 51$$

5. Member slenderness reduction factor

From Table 4.4, section constant $\alpha_b$ = -0.5
From Figure 4.8, $\alpha_c \approx 0.91$

6. Design member capacity

$\phi N_c = 0.9 \times 0.91 \times 1204 = 986$ kN

**Solution according to BS 5950 Part 1**

1. Dimensions and Properties

$b$ = 200 mm
$d$ = 200 mm
$t$ = 5.0 mm
$A_g$ = 3810 mm$^2$
$L$ = 3600 mm
$r$ = 79 mm
$p_y$ = 355 N/mm$^2$

2. Effective area

$b/t = 200/5 = 40 > (b/t)_{limit}$ of 35.8 given in Table 4.1
Flange is not fully effective

$d/t = 200/5 = 40 > (b/t)_{limit}$ of 35.8 given in Table 4.1
Web is not fully effective

From Equation (4.9)

$$A_{eff} = 4\cdot t^2\cdot\left\{\left(\frac{b}{t}\right)_{limit} - 5 + \pi\right\} = 4\cdot 5^2\cdot\{35.8 - 5 + \pi\} = 3394 \text{ mm}^2$$

The flange width to thickness ratio is

$$\frac{b-5\cdot t}{t}=\frac{200-5\times 5}{5}=35$$

The limiting width-to-thickness ratio for class 3 is

$$35\cdot\varepsilon=35\cdot\sqrt{\frac{275}{p_y}}=35\cdot\sqrt{\frac{275}{355}}=30.8$$

Therefore the SHS is a class 4 section.

3. Member slenderness

$k_e$ = 1.0 because of pin-ended condition
For class 4 section, reduced slenderness

$$\lambda\cdot\sqrt{\frac{A_{eff}}{A_g}}=\frac{k_e\cdot L}{r}\cdot\sqrt{\frac{A_{eff}}{A_g}}=\frac{1.0\cdot 3600}{79}\cdot\sqrt{\frac{3394}{3810}}=43$$

4. Compressive strength

From the curve for $p_y$ of 355 N/mm$^2$ in Figure 4.9,

$$p_{cs}=p_c\approx 294\ \text{N/mm}^2$$

5. Member compression resistance

$P_c = A_{eff}\, p_{cs} = 3394 \times 294 = 997{,}836\text{N} = 998$ kN

## 4.4 References


1. AISC (1999), Load and Resistance Factor Design Specification for Steel Hollow Structural Sections, American Institute of Steel Construction, Chicago, Illinois, USA
2. Beedle, L.S. (1992), Why are Specifications Different?, *Journal of Constructional Steel Research*, 17(1-2), pp 1-30
3. BSI (2000), Structural use of Steelwork in Building, BS 5950, Part 1, British Standard Institution, London, UK
4. EC3 (2003), Eurocode 3: Design of Steel Structures – Part 1.1: General Rules and Rules for Buildings, prEN 1993-1-1:2003, November 2003, European Committee for Standardization, Brussels, Belgium
5. Hancock, G.J. and Zhao, X.L. (1992), Research Into the Strength of Cold-Formed Tubular Sections, *Journal of Constructional Steel Research*, 23(1-3), pp 55-72
6. Johnston, B.G. (1976), Guide to Stability Design for Metal Structures, 3rd Edition, Wiley-Interscience, New York, USA

7. Key, P.W. (1988), The Behaviour of Cold-Formed Square Hollow Section Columns, *PhD Thesis*, The University of Sydney, Sydney, Australia
8. Rondal, J., Wurker, K.G., Dutta, D., Wardenier, J. and Yeomans, N. (1992), *Structural Stability of Hollow Sections*, TÜV-Verlag, Köln, Germany
9. Rotter, J.M. , (1982), Multiple Column Curves by Modifying Factors, *Journal of the Structural Division*, ASCE, 108(ST7), pp1665-1669
10. Standards Australia (1998), Steel Structures, Australian Standard AS 4100, Standards Australia, Sydney, Australia
11. Sherman, D. (1992), Tubular Members, In: *Constructional Steel Design: An International Guide*, Dowling, P.J., Harding, J.E. and Bjorhovde, R. (eds), Elsevier Applied Science: London, UK, pp 91-104
12. SSRC (1991), Stability of Metal Structures – A World View, $2^{nd}$ Edition, Structural Stability Research Council, USA
13. Trahair, N.S. and Bradford, M.A. (1988), *The Behaviour and Design of Steel Structures*, Chapman and Hall, London, UK
14. Zhao, X.L., Lim, P., Joseph, P. and Pi, Y.L. (2000), Member Capacity of Columns with Semi-Rigid End Conditions in Oktalok Space Frames, *Structural Engineering and Mechanics – An International Journal*, 10(1), pp 27-36
15. Zhao, X.L. and Grzebieta, R.H. (2002), Strength and Ductility of Concrete Filled Double Skin (SHS inner and SHS outer) Tubes, *Thin-Walled Structures*, 40(2), pp 193-213
16. Zhao, X.L., Grzebieta, R.H. and Elchalakani, M. (2002), Tests of Concrete-Filled Double Skin CHS Composite Stub Columns, *Steel & Composite – An International Journal*, 2(2), pp 129-146

# Chapter 5: Members Subjected to Bending and Compression

## 5.1 Introduction

### *5.1.1 Hollow Sections in Bending and Compression Applications*

Hollow sections are most likely to experience combined bending and compression in low rise portal frame structures. In rigid frame action both the beams and columns experience bending. The beams are subjected to almost zero net axial force, and the columns generally experience relatively low axial compression loads. Rigid jointed trusses such as Vierendeel trusses would also experience combined actions, in which the level of axial compression is generally greater than a portal frame structure.

Members subjected to combined bending and compression are sometimes referred to as "beam-columns" representing the two types of design actions they are intended to resist.

### *5.1.2 Fundamental Behaviour Under Bending and Compression*

The behaviour and strength of a member under bending and compression is related to the strength under bending alone and compression alone with three key modifications:

- Second order effects which possibly magnify the bending moment should be considered in the structural analysis.
- The classification with respect to local buckling may change due to the different stress and strain distributions within the flanges and webs.
- The strength under the combined actions of bending and compression are related to the separate bending and compression strengths via an interaction formula.

This chapter makes reference to the behaviour and strength of hollow sections under pure bending (Chapter 3) and pure compression (Chapter 4).

## 5.2 Second Order Effects

The combination of axial compression and the displaced configuration of a structure can lead to additional bending moments. In the frame shown in Figure 5.1 the relative sway of the ends of a member gives rise to additional second-order moments. This is often called the $P$-$\Delta$ effect. The columns of the portal frame are also subjected to bending and so, in the deflected position, they have a curvature relative to the chord line joining the ends of the member. This also creates second order moments and is commonly termed the $P$-$\delta$ effect. The bending moments in the frame determined from a first-order analysis are termed the *primary moments*. The additional moments resulting from the $P$-$\Delta$ effect due to joint sway and the $P$-$\delta$ effect due to member curvature are called the *secondary moments*.

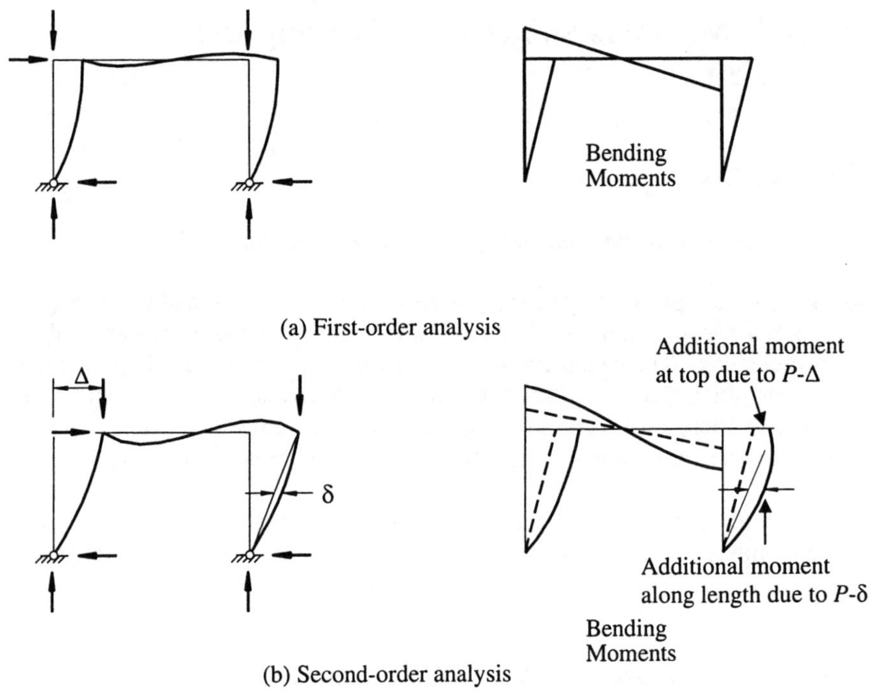


Figure 5.1 First and second-order elastic analysis of a frame.

Second order effects are generally calculated in two ways

- Using a direct second-order elastic analysis. Most modern structural analysis programs now have the ability to perform 2nd order elastic analysis.
- Employing a first-order elastic analysis combined with the approximate determination of second-order moments through *moment amplification factors*.

It is beyond the scope of this book to elaborate on methods of structural analysis, but there are a few important points to note, some of which are particularly relevant to tubular structures:

- Due to the higher strength of many tubular members in use today, smaller, less stiff members are often used. Hence second order effects may be larger in a tubular structure compared to an alternative hot-rolled steel structure. Second-order effects must be considered, although with modern structural analysis programs this is not an imposition on the designer.
- Second order analysis uses an iterative solution procedure, hence designers should be familiar with the convergence parameters, and possible limitations of the software, particularly for unique structures or loading conditions.
- Some structural analysis programs that can perform plastic analyses cannot consider second order effects.

The design equations and examples in this chapter assume that the second order moments have been calculated at all points along a member.

## 5.3 Local Buckling and Section Capacity

### *5.3.1 Additional Effect of Axial Compression Compared to Bending Alone on Local Buckling*

Under bending and compression the stress and strain distribution in a member is altered compared to that for bending only. The position of the neutral axis is changed. Figure 5.2 shows typical idealised stress and strain distributions at first yield and full plasticity for (a) bending alone and (b) combined actions. As the relative level of net axial compression increases, the neutral axis moves further, and a greater proportion of the web experiences compressive stress. Hence the possibility of local buckling in the web increases as the compression increases. This can be considered by reducing the element slenderness limits for the web (previously considered in Chapter 3) as the level of axial compression increases.

The stress distribution in the flange remains generally unchanged in bending and compression compared to bending alone, and hence the flange slenderness limits are not altered for the case of bending and compression.

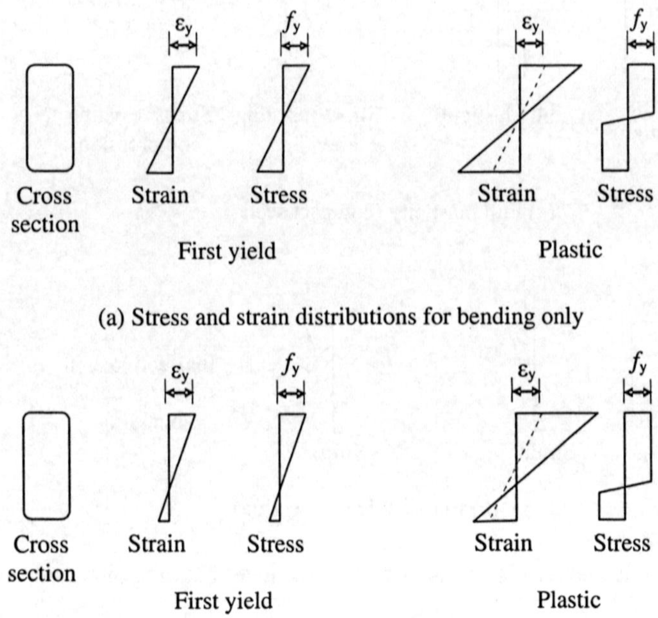


(b) Stress and strain distributions for bending and compression

Figure 5.2 Strain and stress distributions in a hollow section in bending and compression

In addition to the increased likelihood of local buckling, the presence of axial force also affects the amount of bending resistance available from the cross-section compared to bending alone. This is usually considered by calculating a reduced

bending capacity, or using an equivalent interaction equation relating the bending only and compression only strengths.

The stress distributions under bending and compression can be represented in a different manner as shown in Figure 5.3. Under fully yielded conditions, it can be considered that the middle portion of the web, closest to the neutral axis, is providing the resistance against the net axial force, while the flanges and outer portions of the web are providing the bending resistance. Hence, for low levels of axial compression, it is only a small part of the web, very close to the neutral axis that is not resisting bending, and hence the reduction of the bending capacity is almost insignificant.

However, for a Class 3 or Class 4 section under bending and compression, some of the web is ineffective area due to local buckling, and the exact stress distribution is both uncertain and more complicated. Hence the inclusion of additional net axial compression will take away bending resisting areas that are relatively further away from the neutral axis. Therefore the bending capacity of Class 3 and 4 sections reduces faster compared to Class 1 and 2 sections for the same proportion of net axial compression.

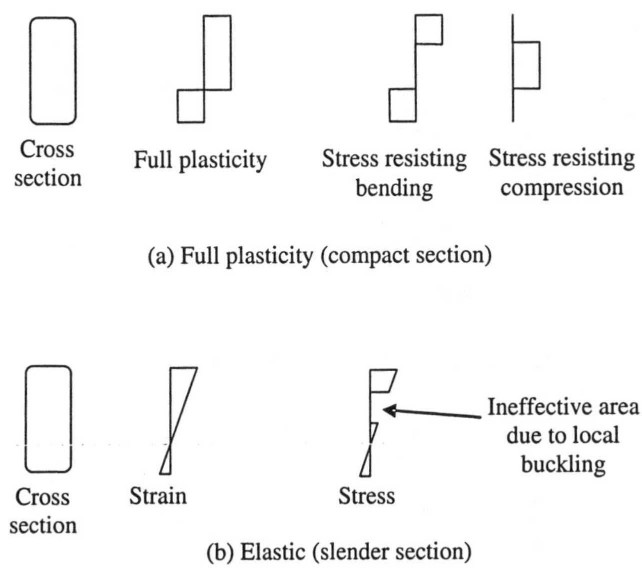


Figure 5.3 Plastic and elastic stress distributions under bending and compression

Figure 5.4 illustrates the typical effect of net axial compression on the bending behaviour of an RHS. Both the overall bending capacity is decreased (though perhaps only slightly) and the rotation capacity is decreased, due to the more severe stress distribution in the web.

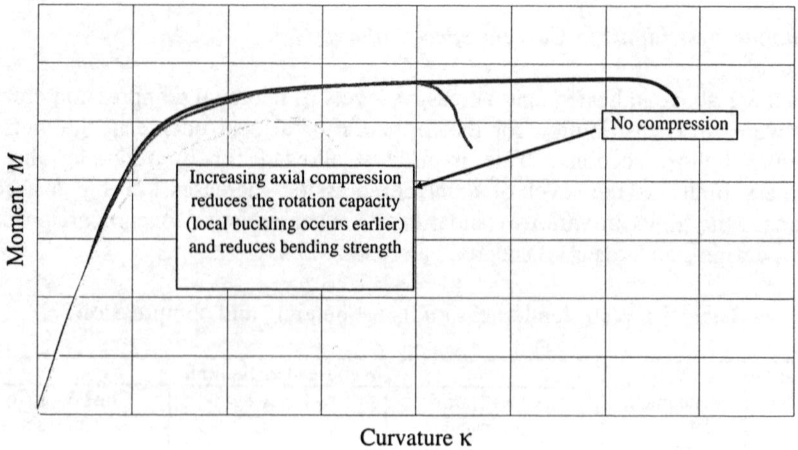


Figure 5.4 Effect of axial compression on the bending behaviour of an RHS – a schematic view

Net axial tension and bending may also occur in tubular members. Using a similar philosophy to above, the slenderness limits for web local buckling would actually increase (though this is often ignored as a conservative assumption), however the bending capacity is decreased in a similar manner as for the case of net axial compression.

The case of bending and axial force of CHS is a more complicated matter, and less design guidance exists on CHS under combined actions. For example, due to the different distribution of area with respect to the neutral axis, net axial force causes a greater reduction in bending capacity compared to an RHS. Conservative assumptions are sometimes applied, hence the more severe slenderness limits for compression only (refer to Chapter 4) can be applied to determine the classification of the section, and the simplest strength interaction equations are usually considered.

The cases of biaxial bending alone, and biaxial bending and net axial force are further complications. For the classification of the section and consideration of local buckling, the stress distributions due to bending about each principal axis are usually considered independently. Interaction for strength can be considered by interaction formulae.

### *5.3.2 Research Basis on Bending and Compression Slenderness Limits*

There have been few investigations of local buckling of hollow sections under bending and compression. Sully (1996) examined the strength equations. Dean and Wilkinson (2001) performed experiments and Dong (2001) performed finite element analysis, but definitive recommendations are still to be published. However it is expected that some reduction in the current slenderness limits will be recommended.

The basis of RHS web slenderness limits is still primarily the bending and compression investigations of I-sections, such as those by Haaijer and Thurlimann (1958) and Dawe and Kulak (1984a, 1984b, 1986).

### *5.3.3 Slenderness Limits in Current Specifications*

Section 5.3.1 above indicates that increasing levels of net axial compression creates a more severe stress distribution for the formation of a local buckle in the web of a rectangular hollow section. This is usually allowed for by reducing the web slenderness limits as the level of axial compression increases. Table 5.1 below summarises the limits in various standards and specifications. A comparison of the different design standards is considered in Section 5.3.5.

Table 5.1 Web slenderness limits for bending and compression

| Standard | Web slenderness definition ($\lambda_w$) | Web slenderness limits | | |
|---|---|---|---|---|
| | | Class 1, or Compact | Class 2 | Class 3, or Non-Compact |
| AS 4100 (See notes (2) and (3)) | $\frac{d-2t}{t}\sqrt{\frac{f_y}{250}}$ | $82.2-137n$ for $n<0.271$ or $52.3-27.4n$ for $n>0.271$ | Not applicable | Not applicable |
| Eurocode 3 (See note (4)) | $\frac{d-2r_{ext}}{t}$ | $\frac{396\varepsilon'}{13\alpha-1}$ for $\alpha>0.5$ or $\frac{36\varepsilon'}{\alpha}$ for $\alpha\le 0.5$ | $\frac{456\varepsilon'}{13\alpha-1}$ for $\alpha>0.5$ or $\frac{41.5\varepsilon'}{\alpha}$ for $\alpha\le 0.5$ | $\frac{42\varepsilon'}{0.67+0.33\psi}$ |
| AISC LRFD HSS (See note (5)) | $\frac{d-2r_{ext}}{t}$ | $3.76\sqrt{\frac{E}{f_y}}(1-2.75n)$ for $n<0.125$ or $1.12\sqrt{\frac{E}{f_y}}(2.33-n)$ for $n>0.125$ | Not applicable | $5.70\sqrt{\frac{E}{f_y}}(1-0.74n)$ |
| BS 5950 (See note (6)) | $\frac{d-5t}{t}$ | $\max\left[\frac{56\varepsilon}{1+0.6r_1}, 35\varepsilon\right]$ | $\max\left[\frac{70\varepsilon}{1+r_1}, 35\varepsilon\right]$ | $\max\left[\frac{105\varepsilon}{1+2r_2}, 35\varepsilon\right]$ |

(1) In all cases $d$ refers to the full depth of the section.
(2) AS 4100 does not specifically prescribe a web slenderness limit in terms of the axial force, but limits the axial force in terms of the web slenderness in Clause 8.4.3.3. This clause only refers to plastic design of doubly symmetric compact I-sections, hence is not strictly applicable to hollow sections.
(3) For AS 4100, $n = N^*/\phi N_s$, the ratio of the design axial force to the design section capacity.
(4) For Eurocode 3, $\alpha$ is the proportion of the web in compression under plastic stress conditions, while $\psi$ is the ratio of the maximum tensile to compressive stress, and $\varepsilon' = \sqrt{(235/p_y)}$.
(5) For AISC LRFD, $n = P_u/\phi_b P_y$, the ratio of the design axial force to the design section capacity.
(6) For BS 5950, $r_1 = F_c/2dtp_{yw}$ (proportion of axial force to yield load of webs) and $r_2 = F_c/A_g p_{yw}$ (proportion of axial force to yield load of the entire section), and $\varepsilon = \sqrt{(275/p_y)}$.

### *5.3.4 Design Rules for Strength – Interaction Formulae*

#### *5.3.4.1 AS 4100*

For SHS and RHS that are compact in bending and where there is no local buckling in compression only (i.e. $k_f$ = 1.0) AS 4100 Clause 8.3.2 gives the reduced nominal section moment capacity ($M_r$) for both major and minor axis bending. This equation also applies to compact SHS and RHS under bending and net axial tension.

$$M_r = 1.18 M_s \left(1 - \frac{N^*}{\phi N_s}\right) \le M_s \tag{5.1}$$

It can be seen that for small net axial force, up to 15 % of the section capacity, there is no reduction in bending capacity.

For SHS and RHS that are compact in bending, where there is local buckling in compression only (i.e. $k_f < 1.0$) AS 4100 Clause 8.3.2 gives the following for both major and minor axis bending.

$$M_r = M_s \left(1 - \frac{N^*}{\phi N_s}\right)\left[1 + 0.18\left(\frac{82 - \lambda_w}{82 - \lambda_{wy}}\right)\right] \le M_s \tag{5.2}$$

where $\lambda_w$ and $\lambda_{wy}$ are the values of $\lambda_e$ and $\lambda_{ey}$ for the web for compression only defined in Chapter 4.

For non-compact or slender SHS and RHS, and all CHS, Clause 8.3.2 of AS 4100 prescribes a simple linear reduction in nominal section moment capacity.

$$M_r = M_s \left(1 - \frac{N^*}{\phi N_s}\right) \tag{5.3}$$

The design section moment capacity is then determined by using the capacity factor $\phi = 0.9$ to give $\phi M_r$. The terms $M_{rx}$ and $\phi M_{rx}$, and $M_{ry}$ and $\phi M_{ry}$ are used when specifically referring to the capacity about either the $x$ or $y$ major or minor principal axes.

For biaxial bending of compact SHS and RHS, AS 4100 Clause 8.3.4 gives the following power law interaction formula:

$$\left(\frac{M_x^*}{\phi M_{rx}}\right)^{\gamma} + \left(\frac{M_y^*}{\phi M_{ry}}\right)^{\gamma} \le 1.0 \text{ where } \gamma = \min\left[1.4 + \left(\frac{N^*}{\phi N_s}\right), 2.0\right] \tag{5.4}$$

For biaxial bending of non-compact and slender SHS and RHS, and all CHS, AS 4100 Clause 8.3.4 gives the following simple linear interaction formula:

$$\frac{N^*}{\phi N_s} + \frac{M_x^*}{\phi M_{sx}} + \frac{M_y^*}{\phi M_{sy}} \le 1.0 \tag{5.5}$$

*5.3.4.2 BS 5950*

The British Standard BS 5950 provides both simple linear interaction strength formulae, and more exact equations for bending and axial force interaction.

Clause 4.8.2.3 and Appendix I.2.2 of BS 5950 suggest that for Class 1 and Class 2 sections, the reduced plastic moment capacities should be calculated "on the basis of the principles of statics" which would involve analysis of the stress distribution in the cross section such as that shown in Figure 5.2(b).

However, the distribution of area for both major and minor axis bending of SHS and RHS is similar to major axis bending for an I-section, except that 2 webs exist in an RHS, so the reduced plastic modulus can be determined by slightly modifying Appendix I.2.1 as follows:

$$S_{rx} = S_x - \left(\frac{A^2}{8t}\right)n^2 \quad \text{for } n = \frac{F}{Ap_y} \le \frac{2t(d-2t)}{A} \tag{5.6a}$$

$$S_{rx} = \left(\frac{A^2}{4b}\right)\left[\left(\frac{2bd}{A}-1\right)+n\right](1-n) \quad \text{for } n = \frac{F}{Ap_y} > \frac{2t(d-2t)}{A} \tag{5.6b}$$

The same equations apply for minor axis bending, though swapping of the dimensions *b* and *d* is required.

A power law interaction formula applies for Class 1 and Class 2 sections for biaxial bending and axial force

$$\left(\frac{M_x}{M_{rx}}\right)^z + \left(\frac{M_y}{M_{ry}}\right)^z \le 1.0 \text{ where } z = \begin{cases} \frac{5}{3}\,(\text{RHS/SHS}) \\ 2.0\,(\text{CHS}) \end{cases} \tag{5.7}$$

For biaxial bending of Class 3 and Class 4 tension members, Clause 4.8.2.2 gives the simplified option of:

$$\frac{F_t}{P_t} + \frac{M_x}{M_{cx}} + \frac{M_y}{M_{cy}} \le 1.0 \tag{5.8}$$

For Class 3 and Class 4 compression members, Clause 4.8.3.2 gives a similar formula

$$\frac{F_c}{A_g p_y} + \frac{M_x}{M_{cx}} + \frac{M_y}{M_{cy}} \le 1.0 \quad (\text{use } A_{eff} \text{ rather than } A_g \text{ for Class 4}) \tag{5.9}$$

Where $M_{cx}$ and $M_{cy}$ are the moment capacities about the *x* and *y* axes.

***5.3.5 Comparison of Specifications***

The various Class 1/Compact web slenderness limits as a function of the changing amount of axial compression are compared in Figure 5.5. Some slight approximations have been made to provide the comparison, such as normalising the web slenderness

with respect to $\sqrt{(f_y/250)}$, or relating the term $r_1$ from BS 5950 to the axial load ratio $n$ in AS 4100.

The slenderness limits for hollow section webs for bending and compression when the compression is zero are naturally the same as the limits for bending alone. Hence the starting point ($n$ = 0) of the relationships in Figure 5.5 represent the bending only limits. The comparison of these values has already been discussed in Section 3.2.6. In this chapter it is sufficient to compare the rate of decrease of the limits with respect to increasing axial force.

AS 4100 and AISC LRFD use bilinear relationships while Eurocode 3 and BS 5950 have curved relationships. All have the similar property that reduction in the limit is more pronounced at low levels of axial force. Apart from the AISC LRFD limit, the remaining 3 specifications merge toward similar values as the level of axial force increases. In many practical situations however, the level of axial force in hollow section beam columns is generally low.

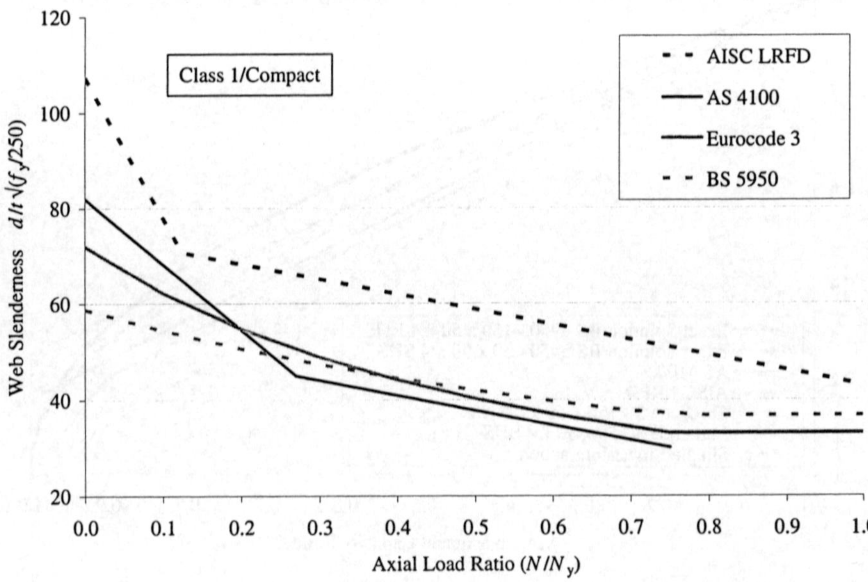


Figure 5.5 Comparison of web slenderness limits with increasing axial compression

The strength interaction formulae for bending and compression are compared in Figure 5.6. Several points can be noted:

- The exact solution depends on the geometry of the cross-section. As the aspect ($d/b$) ratio of the section changes, the relative areas of the flange and web change. For example 50 % compression is required to fully plastify the web of a square hollow section, whereas 75 % compression is required to fully plastify a 3:1 rectangular hollow section. The BS 5950 and Eurocode 3 formulae account for different geometries, whereas AS 4100 and AISC LRFD formulae do not, and must therefore, be slightly conservative in comparison.
- The formulae in BS 5950 (Equation (5.6)), which were given for I-sections, are almost exactly correct for SHS & RHS (though there is a slight variation caused by the corner radii).

- AS 4100 is slightly unconservative (3 %) compared to the exact solution at low levels of axial force, though this is not a considerable concern due to the possibility of strain hardening. However AS 4100 does become conservative up to about 15 % for high aspect ratio sections when the axial force is at 50 %.
- Eurocode 3 can be up to 7 % unconservative for lower levels of axial load (approximately 20 %), but can be up 8 % conservative for higher compressions.
- AISC LRFD is always conservative compared to the exact solution, and can be up to 20 % conservative compared to the exact solution when the axial compression ratio is 50 %.
- The variations between the design standards are not particularly significant, given that the more common applications involving hollow sections under bending and compression have relatively low levels of compression. At 10 % axial load, the specifications vary by no more than 5 %, whereas at 20 % axial load the variation is a maximum 10 %.

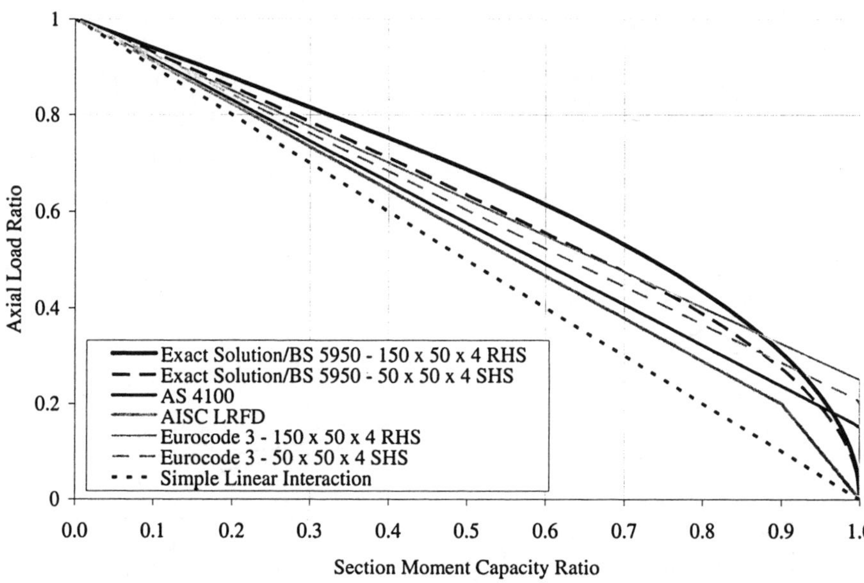


Figure 5.6 Effect of axial load on section moment capacity for major axis bending

### *5.3.6 Examples*

Determine the classification of reduced section moment capacity of a cold-formed C350 (nominal yield stress of 350 N/mm$^2$) RHS 100×50×2.5 for bending about the major principal axis with a design axial compression of 20 kN.

**Solution according to AS 4100**

1. Dimension and Properties

$b = 50$ mm
$d = 100$ mm
$t = 2.5$ mm
$A_n = 709$ mm$^2$
$Z_x = 18.2 \times 10^3$ mm$^3$
$S_x = 22.7 \times 10^3$ mm$^3$
$f_y = 350$ N/mm$^2$
$\phi = 0.9$

Previous calculations
Example 3.2.7 - Bending, $M_s = 7.95$ kNm; $\phi M_s = 7.15$ kNm; compact.
Example 4.2.4 - Compression, $N_s = 231$ kN; $\phi N_s = 208$ kN; not fully effective, $k_f < 1$.

2. Classification

AS 4100 does not specifically prescribe a web slenderness limit in terms of the axial force, but limits the axial force in terms of the web slenderness in Clause 8.4.3.3. This clause only refers to plastic design of doubly symmetric compact I-sections, hence is not strictly applicable to hollow sections, however for the sake of comparison, the calculations are given below.

Design compression, $N^* = 20$ kN
Design section capacity, $\phi N_s = 208$ kN
Axial load ratio, $n = N^*/\phi N_s = 0.0962$

Web slenderness limits for bending only, $\lambda_{ewp} = 82$, $\lambda_{ewy} = 115$
Web slenderness limit reduced for compression, $\lambda_{ewp} = 82.2 - 137n = 69.0$

$$\text{Web slenderness } \lambda_{ew} = \frac{d-2t}{t}\sqrt{\frac{f_y}{250}} = \frac{100-2\times2.5}{2.5}\sqrt{\frac{350}{250}} = 44.96$$

Hence the web is still compact for bending and compression.

3. Strength

The section is compact in bending but not fully effective in compression, so Equation (5.2) (AS 4100 Clause 8.3.2) applies.

$$M_r = M_s\left(1-\frac{N^*}{\phi N_s}\right)\left[1+0.18\left(\frac{82-\lambda_{ew}}{82-\lambda_{ewy}}\right)\right] \le M_s$$
$$= 7.95\left(1-\frac{20}{208}\right)\left[1+0.18\left(\frac{82-44.96}{82-40}\right)\right]$$
$$= 8.32 (> M_s \Rightarrow \text{adopt } M_s)$$
$$= 7.95 \text{ kNm}$$

Design moment capacity, $\phi M_r = 0.9 \times 7.95 = 7.15$ kNm

In this example the relatively small level of axial force (9 %) does not cause any reduction in the section moment capacity. However, it should be remembered that the design bending moment ($M^*$) must include any second order effects.

**Solution according to BS 5950 Part 1**

1. Dimension and Properties

$b = 50$ mm
$d = 100$ mm
$t = 2.5$ mm
$b - 5t = 37.5$ mm
$d - 5t = 87.5$ mm
$A_n = 709$ mm$^2$
$Z_x = 18.2 \times 10^3$ mm$^3$
$S_x = 22.7 \times 10^3$ mm$^3$
$p_y = 350$ N/mm$^2$
$\varepsilon = (275/p_y)^{0.5} = 0.886$

Previous calculations
Example 3.2.7 - Bending, $M_c = 7.95$ kNm; Class 1.
Example 4.2.4 - Compression, $P_c = 229$ kN; not fully effective.

2. Classification

First determine proportion of web in compression
$r_1 = F_c/2(d-5t)tp_{yw} = (20\times10^3)/(2\times87.5\times2.5\times350) = 0.131$

Web slenderness
$(d-5t)/t = 87.5/2.5 = 35$

$$\text{Reduced Class 1 limit} = \max\left[\frac{56\varepsilon}{1+0.6r_1}, 35\varepsilon\right] = \max\left[\frac{56\sqrt{\frac{275}{350}}}{1+0.6\times0.131}, 35\sqrt{\frac{275}{350}}\right] = 46.0$$

The section is still Class 1. In fact, the section could hold approximately 33 kN of axial compression before it changed to a Class 2 section.

3. Strength

Since $n = \dfrac{F}{Ap_y} = \dfrac{20\times10^3}{709\times350} = 0.0806 \le \dfrac{2t(d-2t)}{A} = \dfrac{2\times2.5(100-2\times2.5)}{709} = 0.670$ then the first of the options given in Equation (5.6) should be used.

$$S_{rx} = S_x - \left(\frac{A^2}{8t}\right)n^2$$
$$= 22.7\times10^3 - \left(\frac{709^2}{8\times2.5}\right)0.0806^2$$
$$= 22.54\times10^3 \text{ mm}^3$$

Section capacity is

$M_c = p_y S_{rx} = 350 \times 22.54 \times 10^3 = 7.89$ kNm

For the approximately 9 % level of axial compression there is an almost negligible drop in the section moment capacity of approximately 1 %. However, it should be remembered that the design bending moment must include any second order effects.

## 5.4 Member Buckling and Member Capacity

### *5.4.1 Introduction*

The previous section examined section strength under combined actions and failure generally associated with local buckling. For "long" members, member buckling, related to flexural (Euler) buckling of a column, or flexural torsional buckling of a beam, should be considered.

For a member subjected to axial compression and uniaxial bending about the major principal $x$-axis of the cross-section, the strength of the member may be limited by a an overall in-plane member strength criterion relating to the in-plane bending of beams and flexural buckling of compression members about the major axis. An example of a failed specimen is shown in Figure 5.7. If the member is not completely restrained from deflecting laterally, then it may buckle prematurely out of the plane of bending. The flexural-torsional buckling strength for bending alone is reduced due to possible interaction with flexural (compression) buckling about the minor axis.

For axial compression and uniaxial bending about the minor principal $y$-axis there is no possibility that the member will fail in an out-of-plane mode because it is already deflecting about its weak plane. The strength of the member may be limited by an overall in-plane member strength criterion involving bending and flexural buckling about the minor-axis.

For axial compression and biaxial bending, the failure may be governed by an in-plane member strength criterion, or an out-of-plane member strength criterion.

Figure 5.7 SHS beam - column at failure

### *5.4.2 In Plane Failure*

From the Euler buckling formula for a perfectly elastic column ($N_{euler} = \pi^2 EI/l_e^2$), the capacity of a long column to resist flexural buckling under compression alone is primarily a function of its elastic stiffness ($EI$) and interaction with yielding. Other factors such residual stresses and initial out-of-straightness imperfections also affect the capacity.

For a beam-column, the additional presence of bending moment will cause yielding to occur at lower loads, and since only a proportion of the full cross section is available to resist the axial compression, this will result in a lower effective stiffness to resist flexural buckling under compression. Consequently the capacity is reduced.

Trahair at al (2001) give an in-plane interaction relationship between axial compression and bending. The relationship was developed from both test results and analytical studies on the behaviour of I-sections. Sully (1996) showed that the current interaction design rules in AS 4100 for doubly symmetric compact I-sections are applicable to compact cold-formed RHS. They are conservative for larger values of β (the end moment ratio) and there is scope for improved design rules to be developed.

The behaviour can be approximated by the following equation:

$$\frac{N^*}{N_c} + \frac{M^*}{M_s} \frac{c_m}{1 - N^*/N_{euler}} \leq 1.0 \qquad (5.10)$$

where $N_{euler}$ is the usual elastic flexural buckling load about the applicable axis and $c_m$ is a term to account for the bending moment distribution, and equals 1 for the case

shown with equal end moments (uniform moment distribution). $N_c$ is the member compression capacity while $M_s$ is the bending section capacity.

The interaction between bending moment and compression is given in Figure 5.8 for a selection of different column slenderness ratios (40, 80, 120).

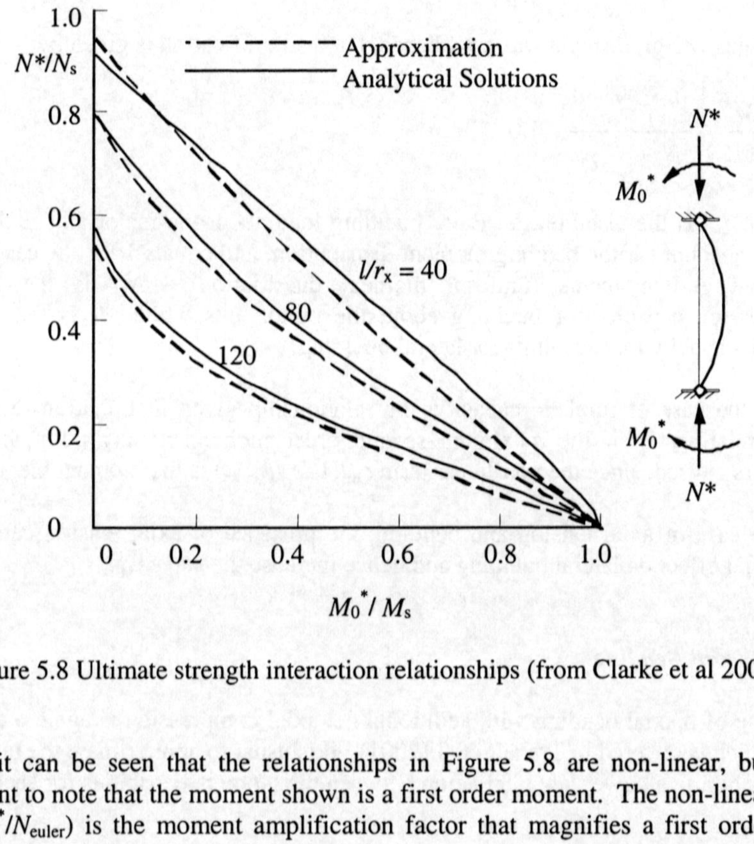


Figure 5.8 Ultimate strength interaction relationships (from Clarke et al 2004)

Firstly it can be seen that the relationships in Figure 5.8 are non-linear, but it is important to note that the moment shown is a first order moment. The non-linear term $c_m/(1-N^*/N_{euler})$ is the moment amplification factor that magnifies a first order to a second order bending moment for a braced member. Hence the approximate solution in Equation (5.10) is in fact linear if the maximum second order moment at any point along the member is used (since the additional non-linear term can then be removed from the equation).

The above behaviour is similar for the case of either major $x$ axis or minor $y$ axis bending.

### *5.4.3 Out of Plane Failure*

A beam bending about its major axis may buckle out of the plane of loading, via a lateral buckle (also called a flexural torsional buckle). This was previously discussed in Section 3.3. Flexural buckling, about the minor axis, caused by axial compression will therefore decrease the out of plane bending capacity of a hollow section. It should be noted that the out-of-plane buckling check only needs to be performed on

RHS bending about the major axis, as SHS and CHS, and RHS bending about the minor axis do not experience lateral buckling under bending.

The analysis of out-of-plane buckling under combined bending and compression is very complex, and beyond the scope of this book. Trahair (1993) give a detailed description of the behaviour of I-section members. It is common to assume that this behaviour of I-sections can be applied to RHS.

A complex design formula was established, but a simplification is given by:

$$\frac{N^*}{N_{cy}}+\frac{M^*}{M_b}\frac{c_m}{1-N^*/N_{x,euler}}\leq 1.0 \qquad (5.11)$$

where $N_{euler}$ is the usual elastic Euler buckling load about the major axis and $c_m$ is a term to account for the bending moment distribution, and equals 1 for the case shown of equal end moments (uniform moment distribution). $N_{cy}$ is the member compression capacity for buckling about the minor axis while $M_b$ is the member bending capacity that accounts for lateral buckling.

As for the case of in-plane capacity, the relationship given in Equation (5.11) is a linear relationship if the maximum second order moment at any point along the members is used, since the non-linear term $c_m/(1-N^*/N_{euler})$ is the moment magnifier.

For the case of axial tension and bending, the presence of axial tension can have a beneficial effect on lateral buckling and hence increase the capacity.

### *5.4.4 Biaxial Bending*

The case of biaxial bending with additional net axial compression is again a complex case which is covered in Trahair et al (2001). It is usual to approximate the behaviour by considering a power law interaction between the moment capacities for each axis.

### *5.4.5 Research Basis on Bending and Compression Slenderness Limits*

There have been few investigations of local buckling of hollow sections under bending and compression with much of the research on slenderness limits being based on investigations of I-sections referred to in Section 3.2.3. Sully (1996) examined the strength equations. Dean and Wilkinson (2001) performed experiments and Dong (2001) performed finite element analysis, but definitive recommendations are still to be published. However it is expected that some reduction in the current slenderness limits will be recommended.

### *5.4.6 Design Rules for Strength*

#### *5.4.6.1 AS 4100*

For in-plane strength, Clause 8.4.2.2 of AS 4100 specifies

$$M_i = M_s\left(1 - \frac{N^*}{\phi N_c}\right) \tag{5.12}$$

The capacities $N_c$ and $M_s$ used in the above equations relate to the plane of bending. Equation (5.12) is equivalent to

$$\frac{N^*}{\phi N_c} + \frac{M^*}{\phi M_s} \leq 1.0 \tag{5.13}$$

A slight complication to Equation (5.12) is the case when the effective length factor for compression is greater than one. For that case, it is sufficient to use $N_c$ in Equation (5.12) based on an effective length factor of one, provided that the compression capacity alone based on the true effective length is also checked. The rationale is that the stability effects have already been considered through the amplification of moments from first-order to second-order values using the amplification factors or a second order analysis. It can be argued that to base the column capacity $N_c$ on the effective length $k_eL$ is to include the stability effects a second time, and that this is inappropriate. This cut-off is demonstrated in Figure 5.9.

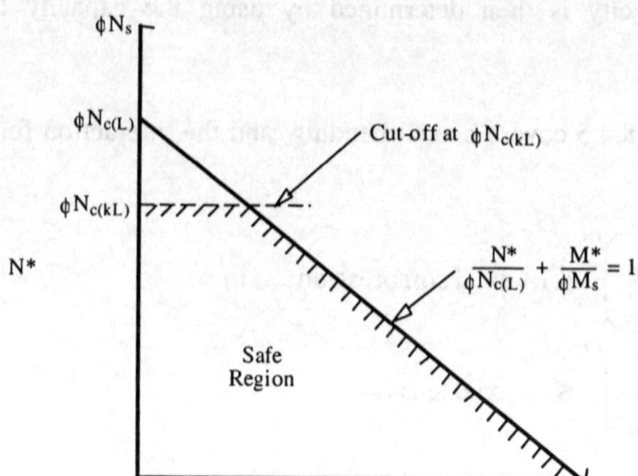


Figure 5.9 AS 4100 beam-column strength equation with cut-off on column strength

Alternatively, Clause 8.4.2.2 of AS 4100 allows a more exact equation for RHS and SHS that are compact and fully effective in compression ($k_f = 1.0$):

$$M_i = M_s\left\{\left[1 - \left(\frac{1+\beta_m}{2}\right)^3\right]\left(1 - \frac{N^*}{\phi N_c}\right) + 1.18\left(\frac{1+\beta_m}{2}\right)^3\sqrt{\left(1 - \frac{N^*}{\phi N_c}\right)}\right\} \leq M_{rx} \text{ or } M_{ry} \tag{5.14}$$

$M_i < M_{rx}$ or $M_{ry}$ depending upon the axis of bending.

where $\beta_m$ is a term accounting for the shape of the bending moment distribution. For the case of end moments only it is the ratio of the smaller to the larger end moments and taken as positive for the case of double (reverse) curvature, and for other cases reference to AS 4100 Clause 4.4.2.2 should be made.

The design capacity is then determined by using the capacity factor $\phi = 0.9$ to give $\phi M_{ix}$ or $\phi M_{iy}$ (whichever is appropriate).

For out-of-plane strength with axial compression, Clause 8.4.4.1 of AS 4100 specifies

$$M_{ox} = M_{bx}\left(1 - \frac{N^*}{\phi N_{cy}}\right) \quad (5.15)$$

which is equivalent to

$$\frac{N^*}{\phi N_{cy}} + \frac{M^*}{\phi M_{bx}} \leq 1.0 \quad (5.16)$$

For out-of-plane strength with axial tension, Clause 8.4.4.2 of AS 4100 specifies

$$M_{ox} = M_{bx}\left(1 + \frac{N^*}{\phi N_t}\right) \leq M_{rx} \quad (5.17)$$

The design capacity is then determined by using the capacity factor $\phi = 0.9$ to give $\phi M_{ox}$.

AS 4100 Clause 8.4.5 covers biaxial bending, and the interaction formulae are given by:

$$\left(\frac{M_x^*}{\phi M_{cx}}\right)^{1.4} + \left(\frac{M_y^*}{\phi M_{iy}}\right)^{1.4} \leq 1 \quad \text{axial compression} \quad (5.18a)$$

$$\left(\frac{M_x^*}{\phi M_{tx}}\right)^{1.4} + \left(\frac{M_y^*}{\phi M_{ry}}\right)^{1.4} \leq 1 \quad \text{axial tension} \quad (5.18b)$$

where

$M_{cx}$ = min[$M_{ix}$, $M_{ox}$] (from Equations (5.12), (5.14), (5.15))

$M_{tx}$ = min[$M_{rx}$, $M_{ox}$] (from Equations (5.1), (5.2), (5.3), (5.17))

*5.4.6.2 BS 5950*

BS 5950 has simplified rules that apply to all sections, and then some more complex, but more exact options that may also be used.

Clause 4.8.3.3 of BS 5959 presents a simplified method:

$$\frac{F_c}{P_c}+\frac{m_x M_x}{p_y Z_x}+\frac{m_y M_y}{p_y Z_y}\le 1.0 \tag{5.19}$$

$$\frac{F_c}{P_{cy}}+\frac{m_{LT} M_{LT}}{M_b}+\frac{m_y M_y}{p_y Z_y}\le 1.0 \tag{5.20}$$

Where

$F_c$ = axial compression

$P_c$ = minimum of major & minor axis compression buckling capacities ($P_{cx}$, $P_{cy}$)

$m_x$, $m_y$, $m_{LT}$ = equivalent uniform moment factor for appropriate bending moment

$M_x$, $M_y$, $M_{LT}$ = maximum moment for the appropriate case

$M_b$ = buckling resistance moment

$Z_x$, $Z_y$ = elastic section modulus

A more exact series of equations for CHS, RHS and SHS is also given by Clause 4.8.3.3.3. The out-of-plane buckling check depends on whether the beam length was short enough to avoid the check for lateral-torsional buckling (Clause 4.3.6.1 referred to in Section 3.3.4.2)

For major axis bending and compression:

$$\frac{F_c}{P_{cx}}+\frac{m_x M_x}{M_{cx}}\left(1+0.5\frac{F_c}{P_{cx}}\right)\le 1.0 \quad \text{(in plane)} \tag{5.21}$$

$$\frac{F_c}{P_{cy}}+0.5\frac{m_{LT} M_{LT}}{M_{cx}}\le 1.0 \quad \text{(out of plane – no LTB check)} \tag{5.22}$$

$$\frac{F_c}{P_{cy}}+\frac{m_{LT} M_{LT}}{M_b}\le 1.0 \quad \text{(out of plane – LTB check)} \tag{5.23}$$

For minor axis bending and compression:

$$\frac{F_c}{P_{cy}}+\frac{m_y M_y}{M_{cy}}\left(1+0.5\frac{F_c}{P_{cy}}\right)\le 1.0 \quad \text{(in plane)} \tag{5.24}$$

$$\frac{F_c}{P_{cx}}+0.5\frac{m_{yx} M_y}{M_{cy}}\le 1.0 \quad \text{(out of plane)} \tag{5.25}$$

For biaxial bending and compression:

$$\frac{F_c}{P_{cx}}+\frac{m_x M_x}{M_{cx}}\left(1+0.5\frac{F_c}{P_{cx}}\right)+0.5\frac{m_{yx} M_y}{M_{cy}}\le 1.0 \quad \text{(in plane)} \tag{5.26}$$

$$\frac{F_c}{P_{cy}}+0.5\frac{m_{LT} M_{LT}}{M_{cx}}+\frac{m_y M_y}{M_{cy}}\left(1+0.5\frac{F_c}{P_{cy}}\right)\le 1.0 \quad \text{(out of plane – no LTB check)} \tag{5.27}$$

$$\frac{F_c}{P_{cy}}+\frac{m_{LT} M_{LT}}{M_b}+\frac{m_y M_y}{M_{cy}}\left(1+0.5\frac{F_c}{P_{cy}}\right)\le 1.0 \quad \text{(out of plane – LTB check)} \tag{5.28}$$

$$\frac{m_x M_x\left(1+0.5\frac{F_c}{P_{cx}}\right)}{M_{cx}\left(1-\frac{F_c}{P_{cx}}\right)}+\frac{m_y M_y\left(1+0.5\frac{F_c}{P_{cy}}\right)}{M_{cy}\left(1-\frac{F_c}{P_{cy}}\right)}\le 1.0 \quad \text{(interactive buckling)} \tag{5.29}$$

Additional alternative equations for combined bending and compression are also given in BS 5950 Annex 1.

### *5.4.7 Comparison of Specifications*

Both AS 4100 and BS 5950 allow for different tiers of strength calculation with simple interaction formulae and more complex equations that result in higher capacities.

Figure 5.10 compares the in-plane buckling interaction formulae of AS 4100 and BS 5950 (Equations (5.12), (5.14) and (5.21)). The direct comparison is slightly deceptive, so some points must be considered.

- The higher tier AS 4100 Equation (5.14) has an upper limit of the reduced section moment capacity ($M_{rx}$ or $M_{ry}$) which is not included in the Figure 5.10. In many situations this will govern, so the significant difference shown in some cases of Figure 5.10 will be reduced.
- In the common case of a simply supported beam, the ratio $\beta$ = -1, so the AS 4100 higher tier Equation (5.14) is exactly equal to the lower tier Equation (5.12), and much closer to BS 5950 Equation (5.21).
- The difference between AS 4100 and BS 5950 is larger for mid range values of axial force (approximately 50%). In many situations, the axial load ratio is low, and the difference is not as pronounced.

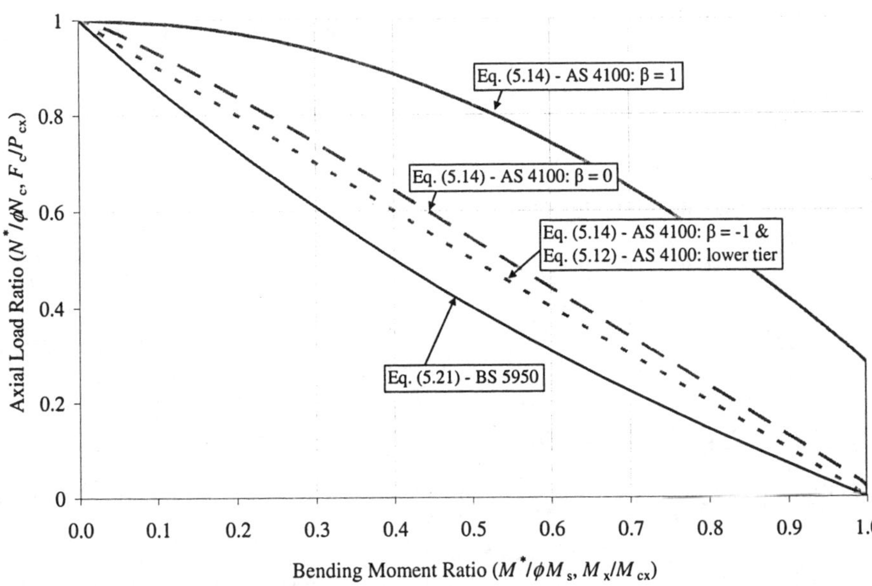


Figure 5.10 Comparison of in-plane interaction formulae

The lower tier options for AS 4100 (Equation (5.12)) and BS 5950 (Equation (5.19)) are similar for the case of uniaxial bending (since the second bending term in Equation (5.19) will become zero). However, the terms in Equation (5.19), $p_yZ_x$ and

$p_y Z_y$ only relate to the yield moments, rather than the actual capacity – and hence are conservative for the Class 1 and Class 2 sections. The case of biaxial bending and compression is different. In the lower tier BS 5950 has a simple linear interaction between the three variables, whereas AS 4100 uses a power law interaction.

For out-of-plane capacity AS 4100 only permits the lower tier simple linear interaction formula (Equation (5.15)). AS 4100 does allow for higher tier formulae to be applied to doubly-symmetric compact I-sections only. BS 5950 allows for higher tier calculations on hollow sections for out of plane buckling (Equations (5.22) and (5.23)) and can produce higher capacities compared to AS 4100.

Section 3.3.2 on lateral buckling referred to a maximum unbraced length (or a maximum member slenderness), below which it could be assumed lateral buckling would not occur for bending moment only. Under combined bending and compression though it is still possible for out of plane buckling to occur. BS 5950 does distinguish between beams which are either side of this limit. The use of Equations (5.22) and (5.23), and (5.27) and (5.28) depends on whether the beam is less than or greater than the critical slenderness. For beams less than the critical slenderness, significant economies can result due to the multiplier of 0.5 in front of the major axis buckling term ($m_{LT}M_{LT}/M_{cx}$). AS 4100 does not have a similar differentiation of these two cases.

### *5.4.8 Examples*

Reconsider Example 3.3.6 which relates to a 100 × 50 × 2.5 C350 RHS which forms a simply supported beam of length 5.3 m from which some roofing and associated services are being suspended. The member only has full restraint at the ends, but no lateral rotation restraints. In addition, the member experiences 20 kN compression (see Figure 5.11). Determine the reduced moment capacity of the beam. It is possible that the beam could buckle either in-plane or out-of-plane. For the purposes of buckling under compression it can be assumed that the effective length factor is one for both directions (ie the effective length is 5.3 m).

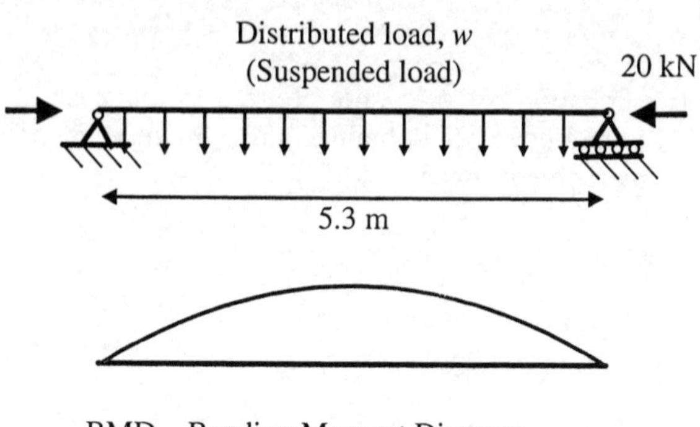


Figure 5.11 Simply supported beam-column

**Solution according to AS 4100**

1. Dimension and Properties

$b = 50$ mm
$d = 100$ mm
$t = 2.5$ mm
$A_n = 709$ mm$^2$
$Z_x = 18.2 \times 10^3$ mm$^3$
$S_x = 22.7 \times 10^3$ mm$^3$
$f_y = 350$ N/mm$^2$
$\phi = 0.9$

Previous calculations
Example 3.2.7 - Bending - local buckling, $M_s = 7.95$ kNm; $\phi M_s = 7.15$ kNm; compact.
Example 3.3.6 - Bending - lateral buckling, $M_b = 7.95$ kNm; $\phi M_{bx} = 7.15$ kNm.
Example 4.2.4 - Compression - local buckling, $N_s = 231$ kN; $\phi N_s = 208$ kN; not fully effective, $k_f < 1$.
Example 4.3.5.1 – Compression – member buckling - $N_{cx} = 60.0$ kN; $\phi N_{cx} = 54.0$ kN;
Example 4.3.5.1 – Compression – member buckling - $N_{cy} = 20.8$ kN; $\phi N_{cy} = 18.7$ kN;
Example 5.3.6 - Bending & Compression - local buckling, $M_{rx} = 7.95$ kN; $\phi M_{rx} = 7.15$ kN.

2. In-Plane Capacity

Since the 100 × 50 × 2.5 C350 RHS was not fully effective in compression, one must use the lower tier Equation (5.12).

$$\begin{aligned}\phi M_{ix} &= \phi M_{sx}\left(1-\frac{N^*}{\phi N_{cx}}\right)\\ &= 7.15\times\left(1-\frac{20}{54.0}\right)\\ &= 4.50 \text{ kNm}\end{aligned}$$

While the higher tier Equation (5.14) cannot be used in this case, if it was, then the same answer for $M_{ix}$ would have been obtained since the moment distribution term is $\beta_m = -1$ for this bending moment shape.

3. Out-of-Plane Capacity

$$\begin{aligned}\phi M_{ox} &= \phi M_{bx}\left(1-\frac{N^*}{\phi N_{cy}}\right)\\ &= 7.15\times\left(1-\frac{20}{18.7}\right)\end{aligned}$$

= undefined as the compression capacity has been exceeded

The compression capacity for minor axis buckling has been exceeded.

The designer decides to insert some bracing to reduce the minor axis effective length to 2.65 m (half the original length).

With the new effective length $\phi N_{cy}$ = 70.7 kN (from Example 4.3.5.1);

$$\begin{aligned}\phi M_{ox} &= \phi M_{bx}\left(1-\frac{N^*}{\phi N_{cy}}\right)\\ &= 7.15\times\left(1-\frac{20}{70.7}\right)\\ &= 5.13\text{ kNm}\end{aligned}$$

4. Second Order Effects

The capacity of the member is governed by in-plane member buckling and the capacity is 4.50 kNm, but this is the maximum 2nd order moment that the beam can resist.

The first order moment ($wL^2/8$) needs to be amplified by the factor
$\delta_b = c_m/(1-N^*/N_{euler}) = 1/(1-[20\times10^3]/[\pi^2 200000\times0.912\times10^6/5300^2]) = 1.45$.

However Clause 4.4.2.1 of AS 4100 states that if $\delta_b > 1.4$, then a proper 2nd order elastic analysis must be carried out.

Using the analysis package Microstran (Engineering Systems 2004) it was found that a uniformly distributed load 0.874 kN/m produced a maximum first order moment of 3.07 kNm and a maximum second order moment of 4.50 kNm. Hence the moment was amplified by a factor of 1.46.

5. Solution

The capacity of the member is governed by in-plane member buckling and the capacity is 4.50 kNm. A second order elastic analysis showed that the maximum uniformly distributed load was 0.874 kN/m.

By comparison, in Example 3.3.6 for the same beam without any axial compression, the design load was 1.90 kN/m.

**Solution according to BS 5950**

1. Dimension and Properties

$b$ = 50 mm
$d$ = 100 mm
$t$ = 2.5 mm
$b - 5t$ = 37.5 mm
$d - 5t$ = 87.5 mm

$A_n = 709\ \text{mm}^2$
$Z_x = 18.2 \times 10^3\ \text{mm}^3$
$S_x = 22.7 \times 10^3\ \text{mm}^3$
$p_y = 350\ \text{N/mm}^2$
$\varepsilon = (275/\,p_y)^{0.5} = 0.886$

Previous calculations
Example 3.2.7 - Bending – local buckling, $M_c = 7.95$ kNm; Class 1.
Example 3.3.6 - Bending – lateral buckling, $M_{cx} = 7.95$ kNm
Example 4.2.4 - Compression, $P_c = 229$ kN; not fully effective.
Example 4.3.5.1 - Compression - member buckling, $p_{cx} = 75$ MPa; $P_{cx} = 53.2$ kN;
Example 4.3.5.1 - Compression - member buckling, $p_{cy} = 29$ MPa; $P_{cy} = 20.5$ kN.

Since the design compression is 20 kN, the axial capacity is only slightly larger (20.5 kN), the designer makes the same decision as was done in the AS 4100 example, to put in some bracing to reduce the effective length for minor axis buckling, and hence increase the capacity.

Changing the effective length to 2.65 m gives
Compression - member buckling, $p_{cy} = 97$ MPa; $P_{cy} = 68.7$ kN (see Example 4.3.5.1).

2. Design for strength

BS 5950 permits the use of different tiers of analysis for this problem. For the sake of comparison, both are considered.

For the case of the parabolic bending moment, the moment modification factor is $m = 0.95$.

$$\frac{F_c}{P_c} + \frac{m_x M_x}{p_y Z_x} + \frac{m_y M_y}{p_y Z_y} = \frac{20}{53.2} + \frac{0.95 M_x}{350 \times 18.2 \times 10^3} + \frac{0}{p_y Z_y} \le 1.0 \Rightarrow M_x = 4.18\ \text{kNm}$$

$$\frac{F_c}{P_{cy}} + \frac{m_{LT} M_{LT}}{M_b} + \frac{m_y M_y}{p_y Z_y} = \frac{20}{68.7} + \frac{0.95 M_x}{7.95} + \frac{0}{p_y Z_y} \le 1.0 \Rightarrow M_x = 5.93\ \text{kNm}$$

However, higher tier options can be used.

$$\frac{F_c}{P_{cx}} + \frac{m_x M_x}{M_{cx}}\left(1 + 0.5\frac{F_c}{P_{cx}}\right) = \frac{20}{53.2} + \frac{0.95 \times M_x}{7.95}\left(1 + 0.5 \times \frac{20}{53.2}\right) \le 1.0 \Rightarrow M_x = 4.40\ \text{kNm}$$

$$\frac{F_c}{P_{cy}} + 0.5\frac{m_{LT} M_{LT}}{M_{cx}} = \frac{20}{68.7} + 0.5 \times \frac{0.95 \times M_{LT}}{7.95} \le 1.0 \Rightarrow M_{LT} = 11.86\ \text{kNm}$$

The second option is giving a value greater than the section capacity, which implies that lateral buckling is not an issue in this case.

3. Second order effects and solution

In plane capacity governs and the capacity is 4.40 kNm. When determining the maximum loading that the member can resist, $2^{nd}$ order effects must be considered. Using the same analysis program, Microstran, as was used in the $2^{nd}$ order analysis for this example according to AS 4100, it was found that a uniformly distributed load of 0.856 kN/m created a second order moment of 4.40 kNm.

**Discussion**

This example highlights some of the differences between the standards identified above.

- For both cases, in plane rather than out-of-plane buckling governed the design once a lateral brace was added at midspan.
- The higher tier design in BS 5950 results higher out-of-plane capacities compared to AS 4100. In particular, since this beam required no lateral buckling check for bending only, the use of the factor 0.5 in Equation (5.22) makes a considerable increase to the out-of-plane capacity.
- The lower tier in-plane buckling formula (Equation (5.19)) was low, since it only uses the yield moment and is hence over conservative for Class 1 or 2 sections.

## 5.5 References

1. AISC (1999), Load and Resistance Factor Design Specification for Structural Steel Buildings, (AISC LRFD), American Institute of Steel Construction, Chicago, Illinois, USA
2. AISC (2000), Load and Resistance Factor Design Specification for Steel Hollow Structural Sections, (AISC LRFD), American Institute of Steel Construction, Chicago, Illinois, USA
3. BSI (2000), Structural use of Steelwork in Building, BS 5950, Part 1, British Standard Institution, London, UK
4. Canadian Standards Association (2001), CAN/CSA-S16-01: Limits States Design of Steel Structures, Toronto, Ontario, Canada
5. Clarke, M.J., Hancock, G.J. and Wilkinson, T. (2004), *Steel Structures 1: Lecture Notes*, Department of Civil Engineering, The University of Sydney, Sydney, Australia
6. Dean M., Wilkinson T. and Hancock G.J. (2001), Bending and Compression Tests of Rectangular Hollow Sections, In: *Tubular Structures IX*, Puthli, R. and Herion, S. (eds), Balkema: Rotterdam, The Netherlands, pp 349-358
7. Dawe, J.L., and Kulak, G.L. (1984a), Plate Instability of W Shapes, *Journal of Structural Engineering*, American Society of Civil Engineers, Vol 110, No 6, June 1984, pp 1278-1291
8. Dawe, J.L., and Kulak, G.L. (1984b), Local Buckling of W Shape Columns and Beams, *Journal of Structural Engineering*, American Society of Civil Engineers, Vol 110, No 6, June 1984, pp 1292-134

9. Dawe, J.L., and Kulak, G.L. (1986), Local Buckling Behaviour of Beam-Columns, *Journal of Structural Engineering*, American Society of Civil Engineers, Vol 112, No 11, November 1986, pp 2447-2461
10. Dong, X. (2001), Finite Element Analysis of Plastic Bending of Cold-Formed Rectangular Hollow Section Beam-Columns, *Masters Thesis*, Department of Civil Engineering, The University of Sydney, Sydney, Australia
11. Engineering Systems (2004), *Microstran*, Structural Analysis Software, http://www.micorostran.com.au
12. EC3 (2003), Eurocode 3: Design of Steel Structures, Part 1-1: General Rules and Rules for Buildings, prEN 1993-1-1:2003, November 2003, European Committee for Standardisation, Brussels, Belgium
13. Haaijer, G. and Thurlimann, B. (1958), On Inelastic Buckling in Steel, *Journal of the Engineering Mechanics Division*, Proceedings of the American Society of Civil Engineers, Vol 84. No EM 2, April 1958, Proceedings Paper No 1581
14. Standards Australia (1998), Australian Standard *AS 4100 Steel Structures*, Standards Australia, Sydney, Australia
15. Sully, R.M. (1996) The Behaviour of Cold-Formed RHS and SHS Beam-Columns, *PhD Thesis*, School of Civil and Mining Engineering, The University of Sydney, Sydney, Australia
16. Trahair, N.S., Bradford, M.A. and Nethercot, D.A. (2001), *The Behaviour and Design of Steel Structures to BS 5950*, Third Edition – British, Spon Press, London, UK
17. Trahair, N.S. (1993), *Flexural-Torsional Buckling of Structures*, E. & F.N. Spon, London, UK

# Chapter 6: Members Subjected to Concentrated Forces

## 6.1 General

Cold-formed RHS members may be subjected to concentrated forces when used in a Vierendeel truss or in a floor system, as shown in Figure 6.1. The concentrated force can be applied to the webs of RHS through a welded brace as in the Vierendeel truss or through a bearing plate as in the floor system. When the force is applied through a welded brace, the major design concern is web buckling and chord flange yielding as explained in Section 6.2. Typical failure modes are shown in Figure 6.2. When the force is applied through a bearing plate, the major design concern is the web bearing capacity and the influence of bending moment in the RHS beam, as explained in Section 6.3. Typical failure modes are shown in Figure 6.3.

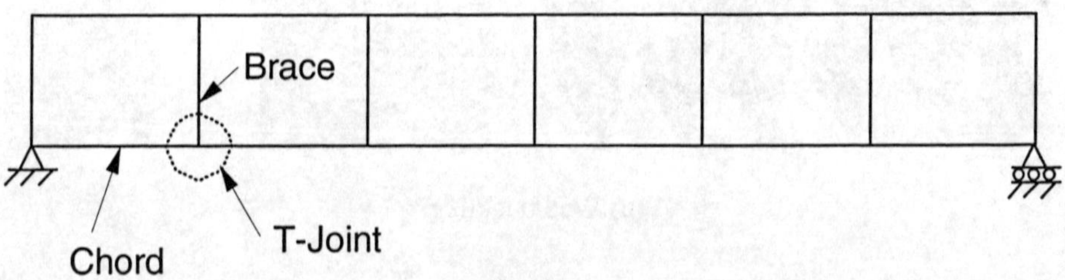


(a) Vierendeel truss (between chord and brace)

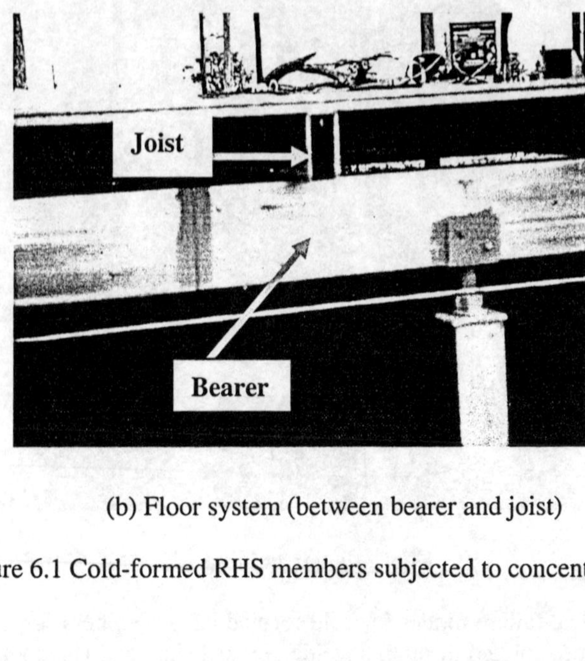


(b) Floor system (between bearer and joist)

Figure 6.1 Cold-formed RHS members subjected to concentrated forces

(a) Web buckling

(b) Chord flange yielding

Figure 6.2 Typical failure modes for cold-formed RHS members subjected to concentrated forces applied through a welded brace (Zhao and Hancock 1991a)

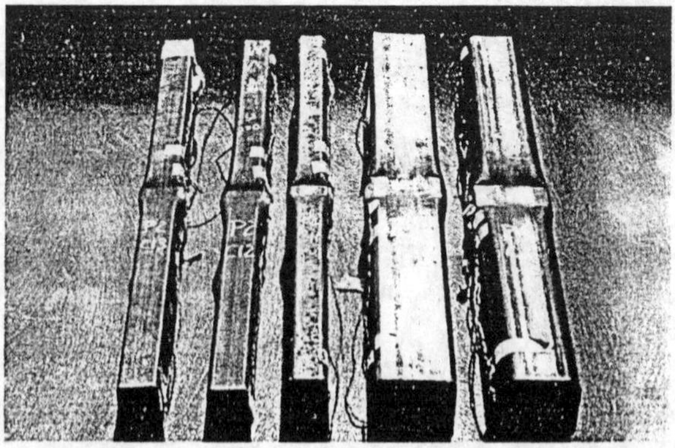

(a) Interior bearing (Zhao 1992)

(b) End bearing (Zhao et al 1996)

(c) Interaction of bending and bearing (Zhao 1992)

Figure 6.3 Typical failure modes for cold-formed RHS members subjected to concentrated forces applied through a bearing plate

## 6.2 Concentrated Forces Applied through a Welded Brace

### *6.2.1 Flange Yielding Versus Web Buckling*

An RHS member subjected to a concentrated force applied through a welded brace (also called branch) is illustrated in Figure 6.4 where various dimensions are defined. The external corner radius ($r_{ext}$) is an important variable in design which can be estimated (ASI 1999) as

$$r_{ext} = 2.5 \cdot t_0 \quad \text{for } t_0 \geq 3 \text{ mm} \tag{6.1a}$$

$$r_{ext} = 2.0 \cdot t_0 \quad \text{for } t_0 < 3 \text{ mm} \tag{6.1b}$$

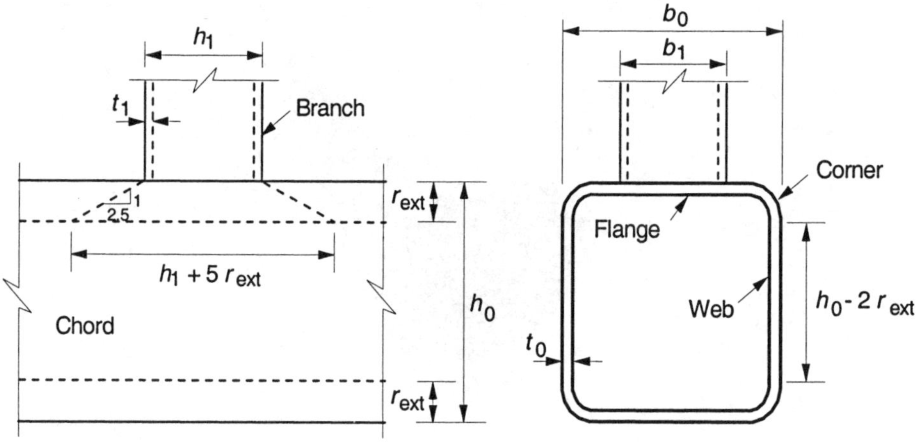


Figure 6.4 An RHS subjected to a concentrated force applied through a welded brace

Tests on T-joints in cold-formed RHS sections were reported by Kato and Nishiyama in Japan (1979) and Zhao and Hancock (1991a) in Australia. There are three main failure modes, namely web buckling failure, chord flange failure and brace local buckling failure. The failure mode of brace local buckling is similar to that observed in a stub column test. This failure mode is not discussed in this chapter since the local buckling can be prevented by using a plate slenderness which is lower than the plate yield slenderness limit of RHS sections as discussed in Chapter 4. A clear peak load is normally found during testing for a web buckling failure mode. The chord flange failure usually has a post-yield response due to the effect of membrane forces in the chord and strain hardening of the material (Zhao and Hancock 1991b). The most important parameter governing the concentrated load behaviour of an RHS T-joint is the ratio of the brace width to the chord width $\beta = b_1/b_0$. This is demonstrated in Figure 6.5 where the failure modes in the test data of Kato and Nishiyama (1980) and Zhao and Hancock (1991a) are presented. It seems that chord flange yielding occurs when $\beta \leq 0.7$. The web buckling failure mode dominates when $\beta \geq 0.8$. The combined failure mode of web buckling and chord flange yielding occurs when $0.7 < \beta < 0.8$. There seems no correlation between the failure modes and web slenderness ratio ($h_0/t_0$).

Based on the research by Zhao (2000) web buckling strength given in Section 6.2.2 can be used for RHS T-joints with $\beta \geq 0.8$ whereas chord flange yielding strength given in Section 6.2.3 can be used for RHS T-joints with $\beta < 0.8$.

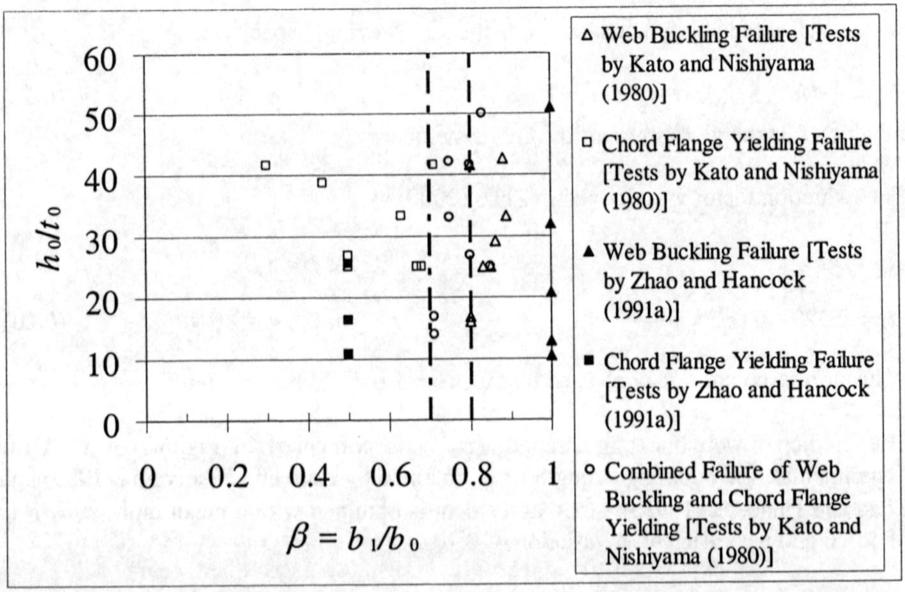


Figure 6.5 Failure modes versus $\beta$ and $h_0/t_0$ for cold-formed RHS T-joints

### 6.2.2 *Ultimate Strength of Web Buckling (for RHS T-joints with $\beta \geq 0.8$)*

Formulae for web buckling of RHS T-joints were previously given in the literature (Packer 1984, Packer et al 1992 and Zhang et al 1989). The predictions of the formulae in these references are compared with the test results on cold-formed RHS T-joints in Zhao (2000). It was found that the CIDECT formula (Packer et al 1992) underestimates the web buckling strength. The formula by Zhang et al (1989) underestimates the web buckling strength for $\beta = 1.0$ and overestimates the web buckling strength for $\beta < 1.0$. The formula by Packer (1984) gives the best prediction with a mean ratio of test to predicted strength of 1.056 with a coefficient of variation of 0.181. The following aspects were not considered in the formulae in the literature to 1992:

(i) Rounded corners of cold-formed RHS sections in calculating the flat web depth.
(ii) Effect of $\beta$ ratio which represents, to some extent, the influence of load eccentricity
(iii) Effect of $(h_0 - 2r_{ext})/t_0$ which represents, to some extent, the influence of column slenderness

The web buckling of cold-formed RHS sections is now treated as a column buckling problem based on the research of Zhao (2000). The column length is assumed to be $(h_0 - 2\ r_{ext})$ where $h_0$ is the overall depth of chord member and $r_{ext}$ is the external

corner radius (see Figure 6.4). The column area is $(h_1 + 5r_{ext}) \cdot t_0$ where $h_1$ is the overall depth of the brace member and $t_0$ is the web thickness as shown in Figure 6.4. The column buckling strength can be expressed simply as:

$$P_{\text{web buckling}} = \alpha_c \cdot N_s \qquad (6.2a)$$

where $\alpha_c$ is a reduction factor and $N_s$ is the section yield capacity, i.e.

$$N_s = 2 \cdot (h_1 + 5 \cdot r_{ext}) \cdot t_0 \cdot f_y \qquad (6.2b)$$

in which $f_y$ is the yield stress of the chord member.

The reduction factor was derived in Zhao (2000) as:

$$\alpha_c = 0.7 \qquad \text{for } \beta = 1.0 \qquad (6.2c)$$

$$\alpha_c = 0.529 - 0.0054 \cdot \frac{h_0 - 2 \cdot r_{ext}}{t_0} \qquad \text{for } 0.8 < \beta \leq 0.9 \qquad (6.2d)$$

A linear interpolation may be used for $0.9 < \beta < 1.0$

The predicted web buckling strength ($\alpha_c N_s$) is compared in Figure 6.6 with the experimental web buckling strength ($P_{ult}$) obtained by Kato and Nishiyama (1980) and Zhao and Hancock (1991a). Good agreement is obtained with a mean ratio ($\alpha_c N_s / P_{ult}$) of 1.026 and a coefficient of variation of 0.102.

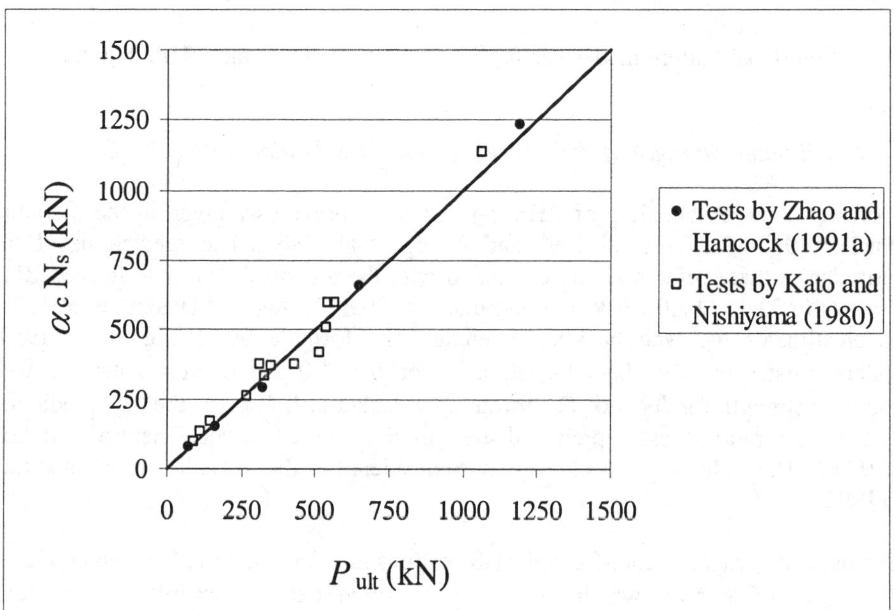


Figure 6.6 Proposed web buckling strength ($\alpha_c N_s$) versus experimental ultimate strength ($P_{ult}$)

### *6.2.3 Ultimate Strength of Chord Flange Yielding (for RHS T-joints with $\beta$ < 0.8)*

Several formulae exist in the literature for chord flange failure of RHS T-joints such as the CIDECT model (Packer et al 1992), the Kato model (Kato and Nishiyama 1980), the modified Kato model (Zhao and Hancock 1991a) and the membrane mechanism model (Zhao and Hancock 1991b). All the formulae were based on yield line mechanism analysis. They were summarized in Zhao (2000). It was found in Zhao (2000) that the modified Kato model gives the best prediction of ultimate strength for cold-formed RHS T-joints with $\beta < 0.8$, as shown in Figure 6.7.

The modified Kato model can be expressed as:

$$P_{\text{ModifiedKato}} = \frac{f_y \cdot t_0^2}{C_1} \cdot \left(2 \cdot C_2 + 4 \cdot \sqrt{C_1}\right) \tag{6.3a}$$

$$C_1 = 1 - \frac{b_1 + 2 \cdot t_0}{b_0} \tag{6.3b}$$

$$C_2 = \frac{h_1 + 2 \cdot t_0}{b_0} \tag{6.3c}$$

The positions of plastic hinges in the modified Kato model are at the top of the web adjacent to the corners rather than at the centre of the corners. It seems that the corners of cold-formed RHS sections should be considered in predicting both web buckling strength and chord flange strength.

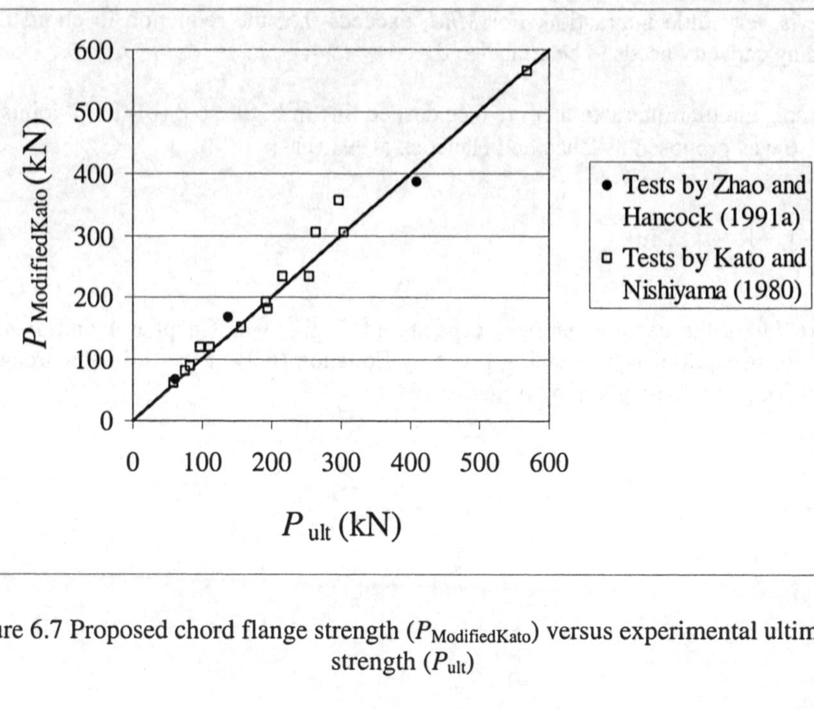


Figure 6.7 Proposed chord flange strength ($P_{\text{ModifiedKato}}$) versus experimental ultimate strength ($P_{\text{ult}}$)

### 6.2.4 *Effect of Bending*

Bending moment often exists in the chord member of a Vierendeel truss. The interaction of bending moment in the chord and the web buckling or chord flange yielding was studied by Zhao and Hancock (1991a). The dimensionless loading ($P/P_f$) versus the dimensionless moment ($M/M_f$) is plotted in Figure 6.8 (a) for $\beta = 1.0$ and in Figure 6.8 (b) for $\beta = 0.5$. The terms $P_f$ and $M_f$ are the test values for the pure concentrated force and pure moment cases respectively. Figure 6.8 (a) corresponds to the effect on web buckling capacity while Figure 6.8 (b) reflects the effect on chord flange yielding capacity.

The increase in capacity above $M$ of $M_f$ in Figure 6.8 (a) is most likely due to the effect of the finite size of the brace member welded to the chord member. For $\beta = 1.0$, increasing the slenderness of the chord has a very little effect on the interaction diagram. It may be concluded that the influence of bending moment on web buckling capacity can be ignored if $P/P_f$ is less than 0.9.

A simple interaction curve (see dashed line in Figure 6.8 (a)) is proposed for T-joints with $\beta = 1.0$ as

$$\frac{P}{P_f} + 0.1 \cdot \frac{M}{M_f} \leq 1.0 \qquad \text{for } P/P_f > 0.9 \qquad (6.4a)$$

$$M \leq M_f \qquad \text{for } P/P_f \leq 0.9 \qquad (6.4b)$$

It seems that the chord flange yielding capacity is influenced by the bending moment in the chord member especially for slender RHS members. For $M/M_f$ less than 0.5, there is very little interaction. For $M/M_f$ exceeds 0.5, the reduction in chord flange yielding capacity needs to be considered.

A simple circular interaction curve (see dashed line in Figure 6.8 (b)) for T-joints with $\beta < 1.0$ was proposed by Zhao and Hancock (1991a) as:

$$\left(\frac{M}{M_f}\right)^2 + \left(\frac{P}{P_f}\right)^2 \leq 1.0 \qquad (6.5)$$

where $M_f$ is the section moment capacity ($M_s$) given in Chapter 4 and $P_f$ is the ultimate strength of web buckling given by Equation (6.2) or the ultimate strength of chord flange yielding given by Equation (6.3).

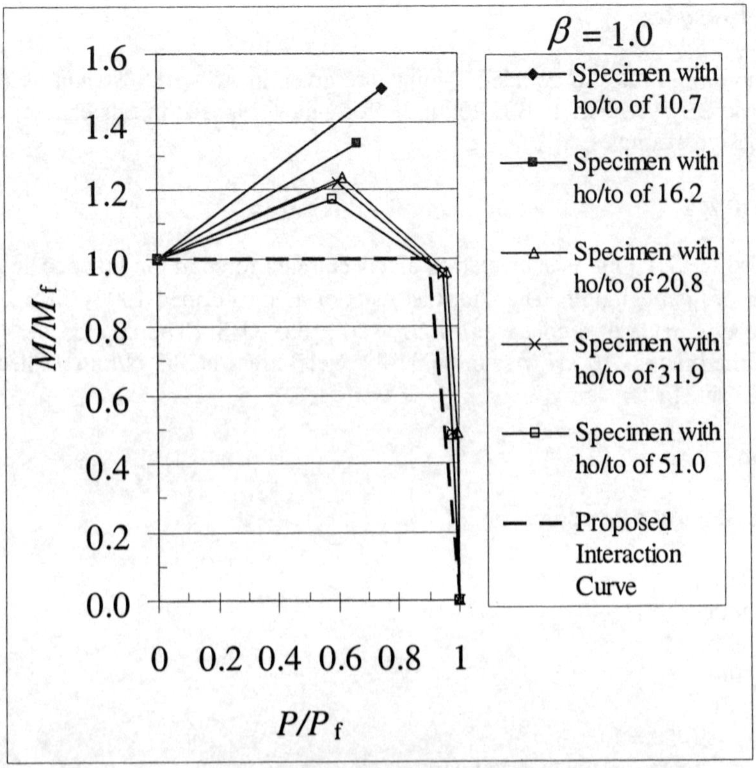


(a) for T-joints with $\beta = 1.0$

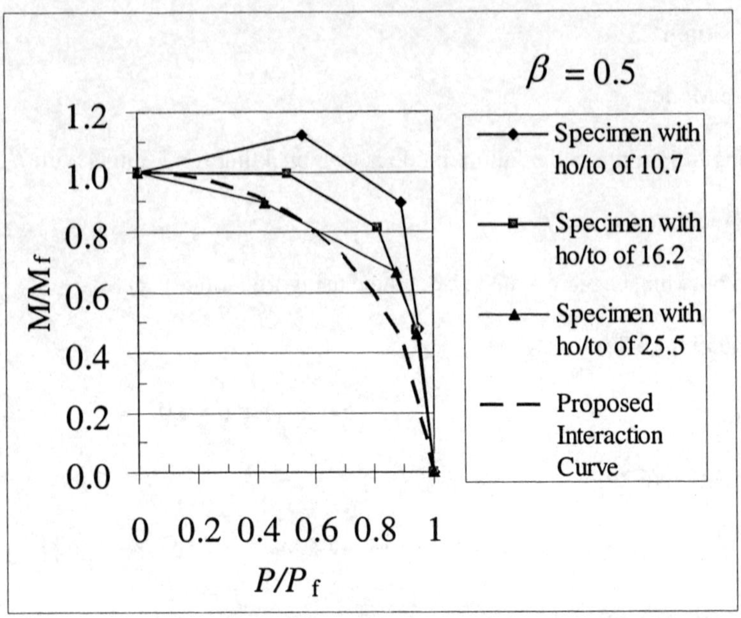


(b) for T-joints with $\beta = 0.5$

Figure 6.8 Interaction diagrams of bending moment and concentrated force based on test data in Zhao and Hancock (1991a)

### 6.2.5 *Examples*

Since no design rules for welded T-joints are given in AS 4100 (Standards Australia 1998) and BS 5950 Part 1 (BSI 2000) the formulae proposed in this chapter are used in the design examples.

#### *6.2.5.1 Example 1*

A welded RHS T-joint is subjected to a concentrated force in the brace. The moment in the chord is negligible. The chord is made of a cold-formed 100 × 50 × 2.5 RHS. The brace is made of a cold-formed 50 × 50 × 5.0 RHS. The applied design force ($R^*$) in the brace is 10 kN. Assume that the yield stress is 355 N/mm$^2$. Check if the T-joint is satisfactory.

**Solution**

1. Dimensions and Properties

$h_0 = 100$ mm
$b_0 = 50$ mm
$t_0 = 2.5$ mm
$h_1 = 50$ mm
$b_1 = 50$ mm
$\beta = b_1/b_0 = 50/50 = 1.0$
From Equation (6.1b) the external corner radius of the chord
$r_{ext} = 2\,t_0 = 2 \times 2.5 = 5$ mm

$f_y = 355$ N/mm$^2$

2. Failure Mode

According to Figure 6.5 the failure mode is web buckling for T-joints with $\beta$ of 1.0.

3. Web Buckling Capacity

The web buckling capacity can be calculated using Equation (6.2).

From Equation (6.2c)

$\alpha_c = 0.7$ for $\beta = 1.0$

From Equation (6.2b)

$$N_s = 2 \cdot (h_1 + 5 \cdot r_{ext}) \cdot t_0 \cdot f_y = 2 \cdot (50 + 5 \cdot 5) \cdot 2.5 \times 355 = 133{,}125 \text{ N} = 133 \text{ kN}$$

From Equation (6.2a)

$$P_{\text{web buckling}} = \alpha_c \cdot N_s = 0.7 \times 133 = 93.1 \text{ kN} \qquad > R^* = 10 \text{ kN}$$

Therefore, the T-joint is satisfactory.

*6.2.5.2 Example 2*

The simply supported beam (see Figure 6.9) has two concentrated loads ($R^* = 10$kN) applied in the same way as described in Section 6.2.5.1 Example 1, i.e. through a welded brace. Full lateral restraint is applied at the location of the loads. The beam is a cold-formed 100 × 50 × 2.5 RHS. The brace is made of a cold-formed 50 × 50 × 5.0 RHS. Assume that the yield stress is 355 N/mm$^2$.

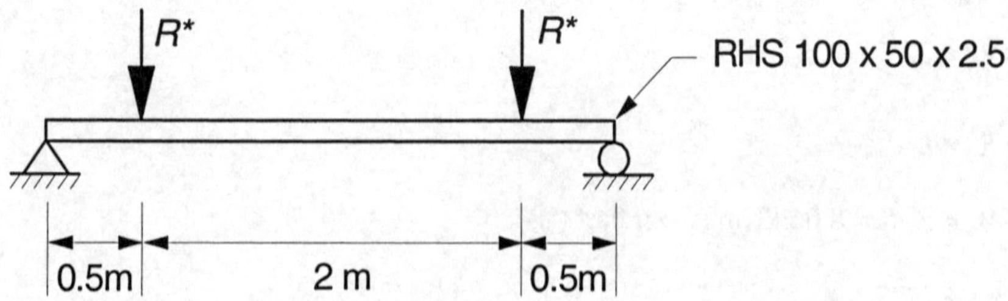


Figure 6.9 A beam under two concentrated forces applied through a welded brace

(i) Check if the T-joint is adequate when the distance between the applied $R^*$ and the end support is 0.5 m.

(ii) If the two loads ($R^*$) move towards each other, what is the maximum distance ($L$) between the applied load $R^*$ and the end support before failure occurs?

**Solution**

(i) The interaction between the bending and concentrated force should be checked against Equation (6.4) for $\beta = 1.0$

$P = R^* = 10$ kN
$P_f = P_{\text{web buckling}} = 93.1$ kN (from Section 6.2.5.1 Example 1)

$P/P_f = R^*/P_{\text{web buckling}} = 10/93.1 = 0.11 < 0.9$

From Equation (6.4b)

$M^* \leq M_f$

$M^* = R^* \times 0.5 = 10 \text{ kN} \times 0.5 \text{ m} = 5 \text{ kNm}$
$M_f = M_s = Z_e f_y$

For RHS 100 × 50 × 2.5 the effective section modulus $Z_{ex} = 22.7 \times 10^3$ mm$^3$ (from ASI 1999)

$M_s = 22.7 \times 10^3 \times 355 = 8.06 \times 10^6 \text{ Nmm} = 8.06 \text{ kNm}$

The T-joint is adequate since $M^* < M_s$.

(ii) Similar to Part (i)

$P = R^* = 10$ kN
$P_f = P_{\text{web buckling}} = 93.1$ kN (from Section 6.2.5.1 Example 1)

$P/P_f = R^*/P_{\text{web buckling}} = 10/93.1 = 0.11 < 0.9$

From Equation (6.4b)

$M^* \leq M_f$

$M^* = R^* \times L$

$M_f = M_s = Z_e f_y = 8.06$ kNm (from Part (i))

$R^* \times L \leq 8.06$

$L \leq 8.06/R^* = 8.06/10 = 0.806$ m = 806 mm

## 6.3 Concentrated Forces Applied through a Bearing Plate

### *6.3.1 General*

The distance ($b_d$ shown in Figure 6.10) between the bearing plate and the edge of an RHS is used to defined the so called "interior bearing" and "end bearing" cases. Interior bearing is the case when $b_d \geq 1.5\ (d - 2r_{ext})$ whereas end bearing is the case when $b_d$ is less than $1.5\ (d - 2r_{ext})$. These two cases are treated separately in design.

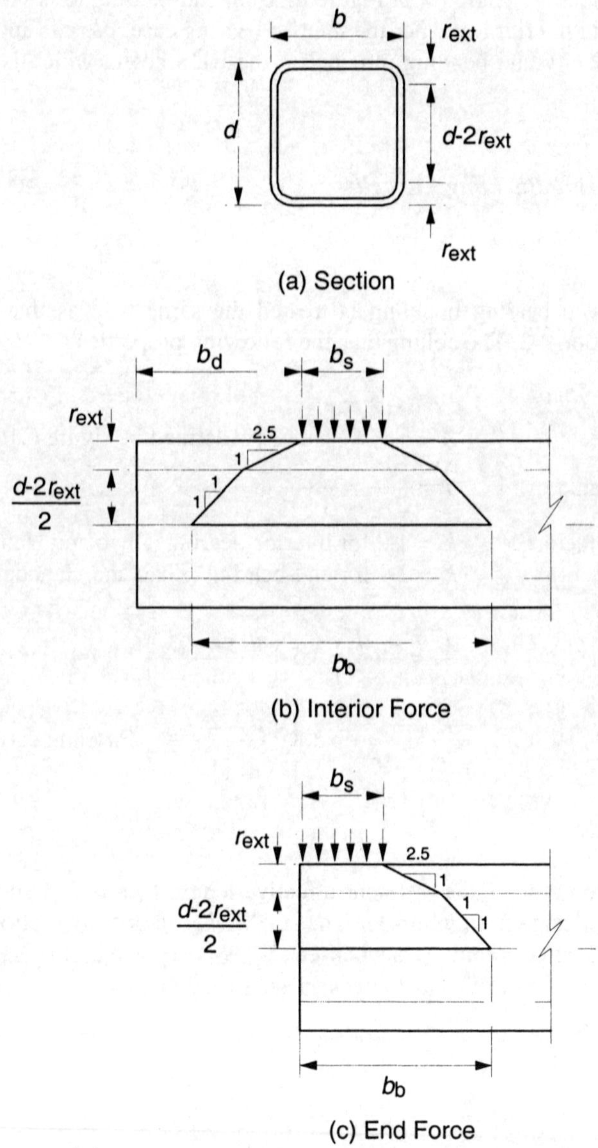


Figure 6.10 Definition of end bearing and interior bearing

Tests on cold-formed RHS subjected to bearing loads were reported by Zhao and Hancock (1992a, 1995). It was found that the external corner radii of RHS introduce load eccentrically to the webs, which is not the case for hot-rolled RHS or I-sections. The eccentric loading produces primary bending of the web out of its plane and a subsequent reduction in capacity. There are two failure modes namely web bearing buckling and web bearing yielding. The web bearing capacity is the lesser of the web bearing buckling capacity and web bearing yield capacity. The most important parameter governing the failure mode is the ratio $(d - 2\,r_{ext})/t$, as shown in Section 6.3.4. The load carrying capacity of RHS under end-bearing force was found to be less on average than half of that of RHS under interior bearing force. The effect of reducing the bearing length ($b_s$ in Figure 6.10) in end-bearing tests was found more severe than that in interior tests. For the interior bearing case, there is an interaction of web bearing capacity and bending moment in the RHS beam, which is addressed in Section 6.3.5.

### 6.3.2 *Web Bearing Buckling Capacity*

#### *6.3.2.1 AS 4100*

The design of web bearing buckling is treated the same way as that of a column described in Section 4.3. The column has the following properties:

Length: $l = d - 2\,r_{ext}$

Cross-section area: $A = b_b \times t$ where $b_b$ is defined in Figure 6.10

Radius of gyration: $r = \dfrac{t}{\sqrt{12}} \approx \dfrac{t}{3.5}$

Effective length factor: $k_e = 1.0$ for interior-bearing (Zhao and Hancock 1992a)
$k_e = 1.1$ for end-bearing (Zhao and Hancock 1995)

Column slenderness:

$$\frac{l_e}{r} = \frac{k_e \cdot l}{r} = 3.5 \cdot \left(\frac{d - 2 \cdot r_{ext}}{t}\right) \text{ for interior bearing}$$

$$\frac{l_e}{r} = \frac{k_e \cdot l}{r} = 3.8 \cdot \left(\frac{d - 2 \cdot r_{ext}}{t}\right) \text{ for end bearing}$$

Form factor: $k_f = 1.0$

It is interesting to compare the different effective length factors ($k_e$) adopted for web bearing buckling design in Figure 6.11. The smaller $k_e$ factor for I-sections (AS 4100) and for Hollow Flange Beams (Hancock et al 1994) represents the larger restraints against rotation provided by two flanges in an I-section or by a closed flange in a hollow flange beam.

The nominal bearing buckling capacity ($R_{bb}$) for an RHS can be expressed as:

$$R_{bb} = 2 \cdot b_b \cdot t \cdot f_y \cdot \alpha_c \quad (6.6)$$

where $t$ is the wall thickness of the RHS, $f_y$ is the yield stress and

$$b_b = b_s + d + 3 \cdot r_{ext} \quad \text{for interior bearing} \quad (6.7a)$$

$$b_b = b_s + 0.5 \cdot d + 1.5 \cdot r_{ext} \quad \text{for end bearing} \quad (6.7b)$$

$\alpha_c$ is the member slenderness reduction factor determined from Equation (4.13) with section constant $\alpha_b = 0.5$ and

$$\lambda_n = 3.5 \cdot \left( \frac{d - 2 \cdot r_{ext}}{t} \right) \cdot \sqrt{\frac{f_y}{250}} \quad \text{for interior bearing} \quad (6.8a)$$

$$\lambda_n = 3.8 \cdot \left( \frac{d - 2 \cdot r_{ext}}{t} \right) \cdot \sqrt{\frac{f_y}{250}} \quad \text{for end bearing} \quad (6.8b)$$

or $\alpha_c$ can be determined using Figure 6.12.

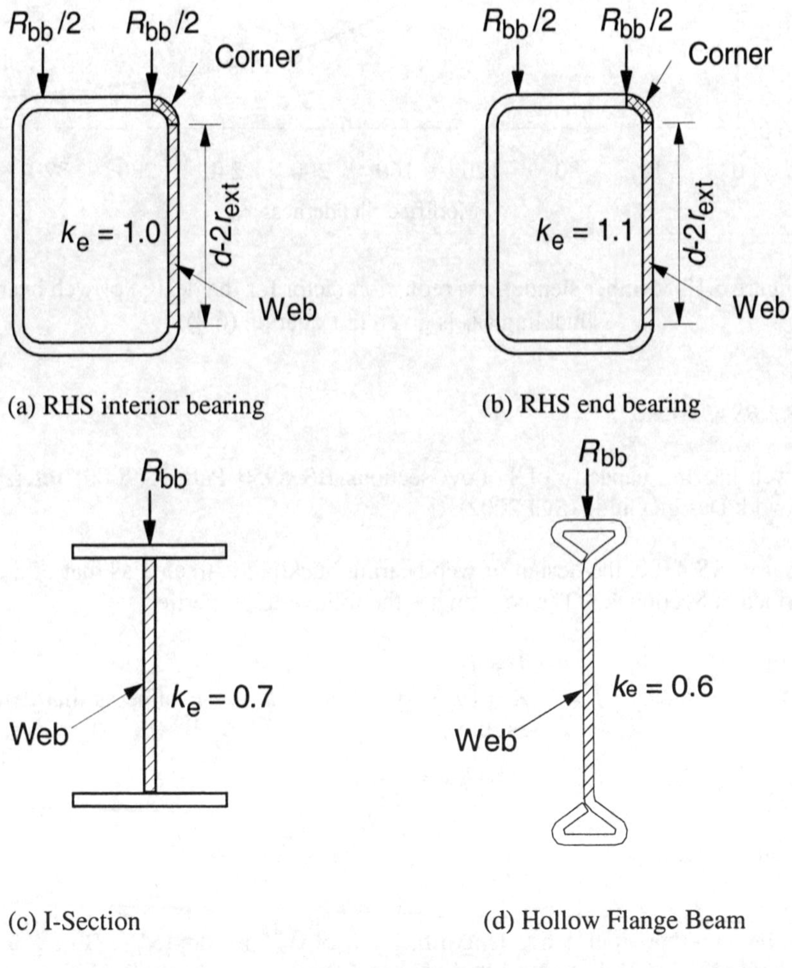


(a) RHS interior bearing (b) RHS end bearing

(c) I-Section (d) Hollow Flange Beam

Figure 6.11 Effective length factors for the design of web bearing buckling

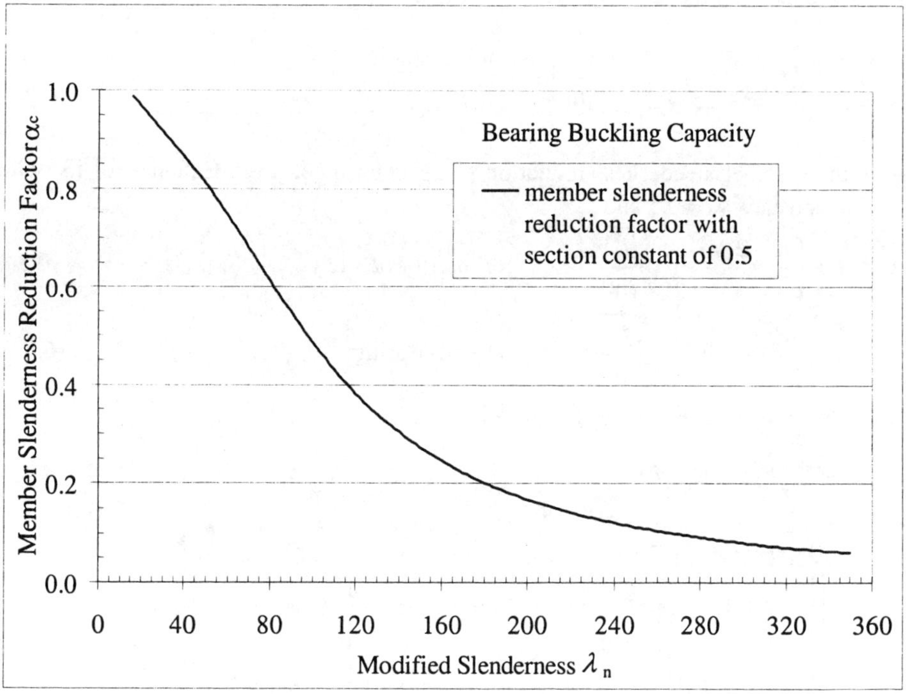


Figure 6.12 Member slenderness reduction factor for the design of web bearing buckling ($\lambda_n$ is given in Equation (6.8))

*6.3.2.2 BS 5950 Part 1*

For web bearing capacity of hollow sections BS 5950 Part 1 (2000) refers to the Steelwork Design Guide (SCI 2002).

Similar to AS 4100, the design of web bearing buckling is treated as that of a column described in Section 4.3. The column has the following properties:

Length: $L = d - 2\,t$

Cross-section area: $A = (b_s + d) \times t$ where $b_s$ is the same as that defined in Figure 6.10

Radius of gyration: $r = \dfrac{t}{\sqrt{12}} \approx \dfrac{t}{3.5}$

Column slenderness: $\dfrac{L_E}{r} = \dfrac{k_e \cdot L}{r} = 0.75 \cdot \sqrt{12} \cdot \left(\dfrac{d - 2 \cdot t}{t}\right)$

It can be seen that an effective length factor $k_e$ of 0.75 is adopted in BS 5950 Part 1. This $k_e$ factor of 0.75 is smaller than that of 1.0 or 1.1 used in AS 4100. However a

shorter length of $d - 2\ r_{ext}$ is used in AS 4100. The effect of load eccentricity due to corner radii is considered in BS 5950 Part 1 as described later.

It should be noted that SCI (2002) does not distinguish end bearing from interior bearing in determining the bearing buckling capacity.

The bearing buckling capacity for an RHS was derived in SCI (2002) where an eccentricity ($e$ in Figure 6.13) is considered. The formulae in SCI can be rewritten in a similar format as that for AS 4100 as:

$$R_{bb} = 2 \cdot (b_s + d) \cdot t \cdot f_y \cdot \alpha_{SCI} \qquad (6.9a)$$

where $b_s$ is the bearing length defined in Figure 6.10, $d$ is the overall depth of an RHS, $t$ is the wall thickness, $f_y$ is the yield stress and $\alpha_{SCI}$ is given by

$$\alpha_{SCI} = \frac{\dfrac{p_c}{f_y}}{1 + K \cdot \dfrac{p_c}{f_y}} \qquad (6.9b)$$

$$K = \left[ \frac{3 \cdot (3 + \frac{d}{b})}{(\frac{d}{b})^2 + 4 \cdot (\frac{d}{b}) + 3} \right] \cdot \left( \frac{e}{t} \right) \cdot \left( \frac{b - t - e}{b - t} \right) \qquad (6.9c)$$

$$e = 0.026 \cdot b + 0.978 \cdot t + 0.002 \cdot d \qquad (6.9d)$$

$p_c$ is the compression strength based on a web slenderness $\lambda = 0.75 \cdot \sqrt{12} \cdot \left( \dfrac{d - 2 \cdot t}{t} \right)$ and strut curve (c), which can be determined using Equation (4.15) or Figure 4.9 for yield stress values of 275 N/mm$^2$, 355 N/mm$^2$ and 460 N/mm$^2$.

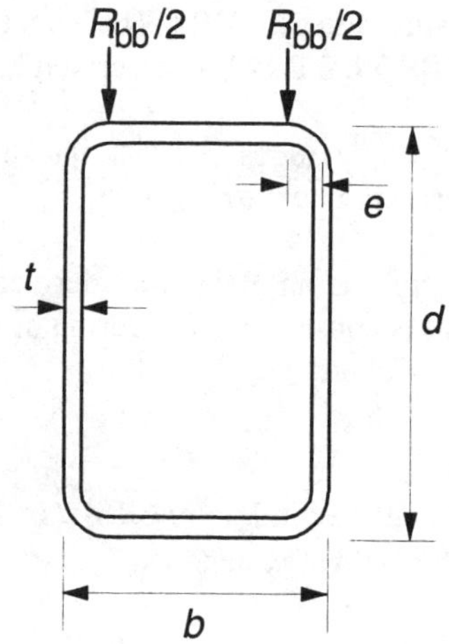


Figure 6.13 Eccentricity ($e$) defined in SCI (2002)

### 6.3.3 *Web Bearing Yield Capacity*

*6.3.3.1 AS 4100*

The design of web bearing yield was based on plastic mechanism analysis shown in Figure 6.14 (a) for interior bearing and Figure 6.14 (b) for end bearing assuming a mechanism length of $b_b$ given in Equation (6.7). The full derivation is given in Zhao and Hancock (1992a, 1995).

The nominal bearing yield capacity ($R_{by}$) for an RHS can be expressed as:

$$R_{by} = 2 \cdot b_b \cdot t \cdot f_y \cdot \alpha_p \quad (6.10a)$$

in which t is the wall thickness of an RHS, $f_y$ is the yield stress, $b_b$ is given in Equation (6.7) and $\alpha_p$ is given by

$$\alpha_p = \frac{0.5}{k_s} \cdot \left[ 1 + (1 - \alpha_{pm}^2) \cdot \left( 1 + \frac{k_s}{k_v} - (1 - \alpha_{pm}^2) \cdot \frac{0.25}{k_v^2} \right) \right] \quad \text{for interior bearing} \quad (6.10b)$$

$$\alpha_p = \sqrt{(2 + k_s^2)} - k_s \quad \text{for end bearing} \quad (6.10c)$$

$$\alpha_{pm} = \frac{1}{k_s} + \frac{0.5}{k_v} \quad (6.10d)$$

$$k_s = \frac{2 \cdot r_{ext}}{t} - 1 \quad (6.10e)$$

$$k_v = \frac{d - 2 \cdot r_{ext}}{t} \quad (6.10f)$$

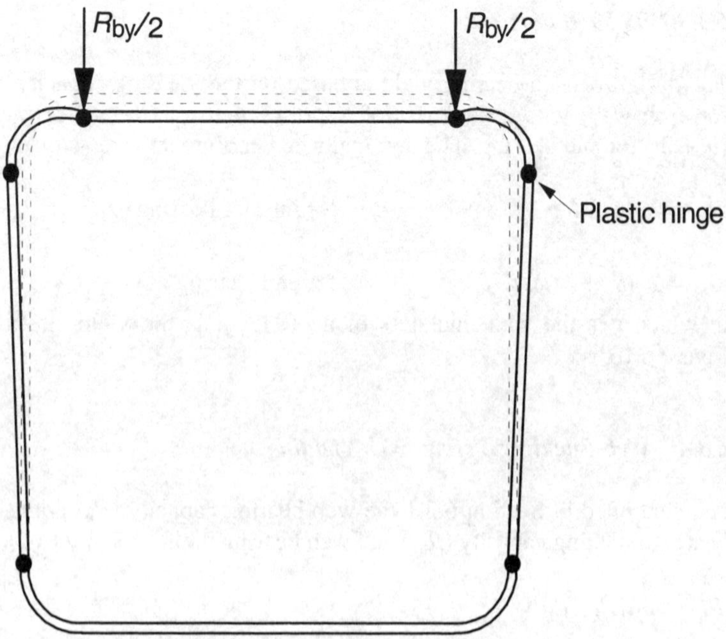


(a) Interior bearing

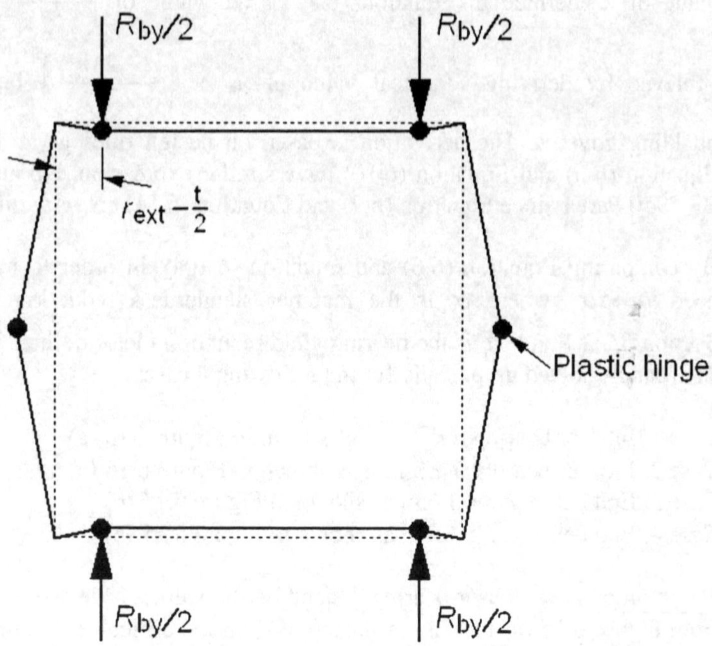


(b) End bearing

Figure 6.14 Mechanism assumed in deriving web bearing yield capacity for RHS

*6.3.3.2 BS 5950 Part 1*

The design of web bearing yield is based on the yielding capacity of two areas (one for each web) with a thickness of $t$ and a length of $(b_s+5t)$ for interior bearing and $(b_s + 2t)$ for end bearing. The nominal web bearing yield capacity can be rewritten as:

$$R_{by} = 2 \cdot (b_s + 5 \cdot t) \cdot t \cdot f_y \qquad \text{for interior bearing} \qquad (6.11a)$$

$$R_{by} = 2 \cdot (b_s + 2 \cdot t) \cdot t \cdot f_y \qquad \text{for end bearing} \qquad (6.11b)$$

in which $t$ is the wall thickness of an RHS, $f_y$ is the yield stress, $b_s$ is defined in Figure 6.10.

### *6.3.4 Web Buckling Versus Web Yielding*

As mentioned in Section 6.3.1 the web bearing capacity ($R_b$) is the lesser of the web bearing buckling capacity ($R_{bb}$) and web bearing yield capacity ($R_{by}$), i.e.

$$R_b = \min\{R_{bb}, R_{by}\} \qquad (6.12)$$

Instead of calculating both $R_{bb}$ and $R_{by}$ every time, it would be helpful for designers if the failure mode (web buckling versus web yielding) can be determined based on the value of $\lambda_n$ defined in Equation (6.8) or the value of $\left(\frac{d - 2 \cdot r_{ext}}{t}\right)$. This can be achieved by deriving a critical value of $\lambda_n$ or $\left(\frac{d - 2 \cdot r_{ext}}{t}\right)$ beyond which web buckling governs. The derivation is based on design rules given in AS 4100 since Equation (6.6) and Equation (6.10) have similar expression. No attempt is made for BS 5950 Part 1 since Equation (6.9) and Equation (6.11) are very different.

By comparing Equation (6.6) and Equation (6.10a), in order to have $R_{bb} < R_{by}$, we need $\alpha_c < \alpha_p$ where $\alpha_c$ is the member slenderness reduction factor defined in Section 6.3.2.1 and $\alpha_p$ is the bearing yield reduction factor defined in Section 6.3.3.1. This can be solved graphically for the following 4 cases.

Case 1: Interior bearing, $t < 3$ mm as shown in Figure 6.15 (a)
Case 2: Interior bearing, $t \geq 3$ mm as shown in Figure 6.15 (a)
Case 3: End bearing, $t < 3$ mm as shown in Figure 6.15 (b)
Case 4: End bearing, $t \geq 3$ mm as shown in Figure 6.15 (b)

The solutions for interior bearing depend on the value of yield stress $f_y$. It can be seen from Figure 6.15 (a) that the influence of $f_y$ on the critical $\lambda_n$ is minimal. The critical values of $\lambda_n$ based on Figure 6.15 are:

$\lambda_{n,critical} = 137$ when $t < 3$mm for both interior bearing and end bearing
$\lambda_{n,critical} = 161$ when $t \geq 3$mm for both interior bearing and end bearing

This can be translated into critical values of $\left(\frac{d-2\cdot r_{ext}}{t}\right)$ as:

$$\left(\frac{d-2\cdot r_{ext}}{t}\right)_{critical}=\frac{\lambda_{n,critical}}{3.5\cdot\sqrt{\frac{f_y}{250}}}\quad \text{for interior bearing} \tag{6.13a}$$

$$\left(\frac{d-2\cdot r_{ext}}{t}\right)_{critical}=\frac{\lambda_{n,critical}}{3.8\cdot\sqrt{\frac{f_y}{250}}}\quad \text{for end bearing} \tag{6.13b}$$

Critical values of $\lambda_n$ and $\left(\frac{d-2\cdot r_{ext}}{t}\right)$ beyond which web buckling governs are listed in Table 6.1 for yield stress values of 250, 275, 350, 355, 450 and 460 N/mm$^2$.

Table 6.1 Critical values of $\lambda_n$ and $\left(\frac{d-2\cdot r_{ext}}{t}\right)$ beyond which web buckling governs

| Bearing type | Thickness ($t$) | Yield stress $f_y$ (MPa) | $\lambda_{n,critical}$ | $\left(\frac{d-2\cdot r_{ext}}{t}\right)_{critical}$ |
|---|---|---|---|---|
| Interior | < 3 mm | 250 | 137 | 39.1 |
| | | 275 | | 37.3 |
| | | 350 | | 33.1 |
| | | 355 | | 32.8 |
| | | 450 | | 29.2 |
| | | 460 | | 28.9 |
| Interior | ≥ 3 mm | 250 | 161 | 46.0 |
| | | 275 | | 43.9 |
| | | 350 | | 38.9 |
| | | 355 | | 38.6 |
| | | 450 | | 34.3 |
| | | 460 | | 33.9 |
| End | < 3 mm | 250 | 137 | 36.1 |
| | | 275 | | 34.4 |
| | | 350 | | 30.5 |
| | | 355 | | 30.3 |
| | | 450 | | 26.9 |
| | | 460 | | 26.6 |
| End | ≥ 3 mm | 250 | 161 | 42.4 |
| | | 275 | | 40.4 |
| | | 350 | | 35.8 |
| | | 355 | | 35.6 |
| | | 450 | | 31.6 |
| | | 460 | | 31.2 |

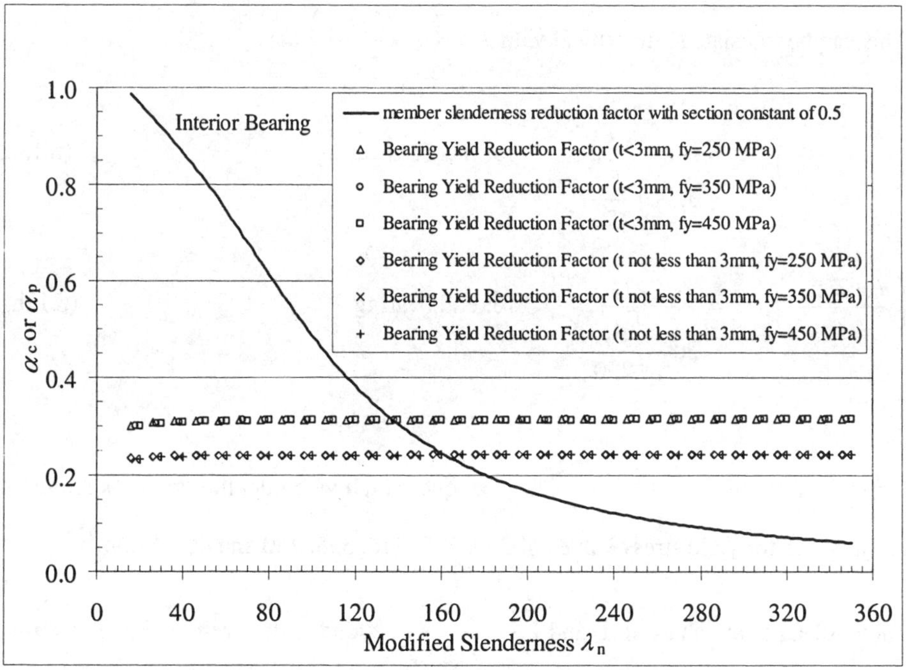


(a) Interior bearing ($\lambda_n$ defined in Equation (6.8a))

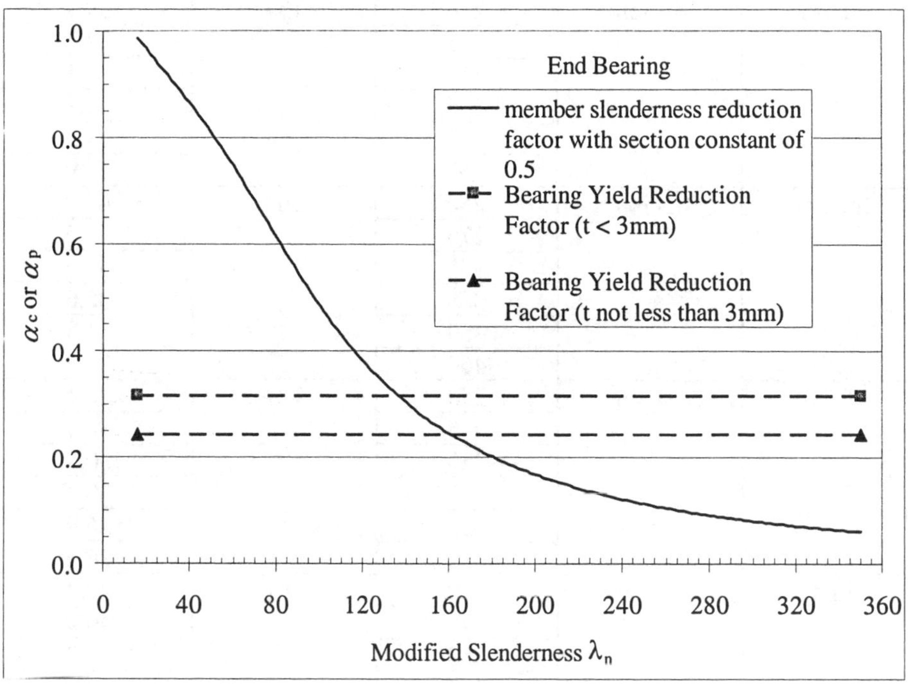


(b) End bearing ($\lambda_n$ defined in Equation (6.8b))

Figure 6.15 Derivations of critical values of $\lambda_n$

### 6.3.5 *Effect of Bending*

The effect of bending on web bearing capacity was studied by Zhao and Hancock (1992a, 1992b). A series of interaction tests of bending and transverse concentrated force were performed. The test results are summarized in Figure 6.16 where those of similar tests on RHS with welded braces (in Figure 6.8 (a)) are also plotted. The interaction is more severe for tests with bearing plate since there is no restraint against web rotation provided by welded braces. It was found that the major parameters affecting the interaction behaviour are the ratio ($\gamma$) of bearing length ($b_s$) to section width ($b$) and the ratio of web depth ($d$ - $2t$) to web thickness ($t$) (Zhao and Hancock 1992a, 1992b).

The interaction equations are given as Equation (6.14) and are also plotted in Figure 6.16:

$$1.2 \cdot \left( \frac{R^*}{\phi \cdot R_b} \right) + \left( \frac{M^*}{\phi \cdot M_s} \right) \leq 1.5 \qquad \text{for } \frac{b_s}{b} \geq 1.0 \text{ and } \frac{d - 2 \cdot t}{t} \leq 30 \qquad (6.14a)$$

or

$$0.8 \cdot \left( \frac{R^*}{\phi \cdot R_b} \right) + \left( \frac{M^*}{\phi \cdot M_s} \right) \leq 1.0 \qquad \text{otherwise} \qquad (6.14b)$$

where $\phi$ (=0.9) is the capacity factor given in Table 3.4 of AS 4100, $R_b$ is the nominal web bearing capacity given in Equation (6.12), $M_s$ is the nominal section moment capacity given in Chapter 3, $b_s$ is the bearing length defined in Figure 6.10, $b$ is the overall width of the RHS and $t$ is the web thickness.

No interaction formulae are given in SCI (2002) for RHS subjected to combined bending and bearing forces. However interaction formulae are given in light gauge cold-formed steel structures codes such as AS/NZS 4600 (Standards Australia 1996), NAS (2002), CSA-S136-01 (2001) and BS 5950 Part 5 (1998) for channel, hat and I-sections. Interaction formulae are also summarised in Trahair et al (2001) for I-section beams. More explanation on this topic can be found in Ratcliff (1975), Hetrakul and Yu (1978, 1979), Wing and Schuster (1986), Rockey et al (1972) and Roberts (1981, 1983). It should be noted that the interaction design rules given in light gauge cold-formed steel structures codes may not be applicable to RHS. This is partly because the normalizing values for the moment and force are those for channel, hat or I-sections, and partly because they may underestimate the interaction for RHS with large ratio of web depth ($d$ - $2t$) to web thickness ($t$).

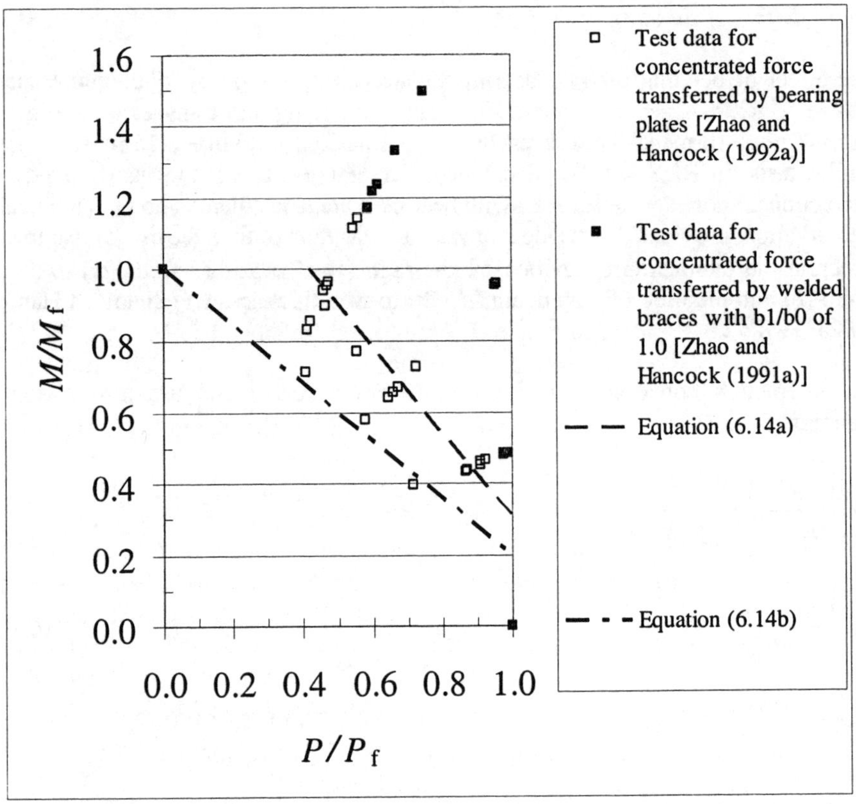


Figure 6.16 Interaction diagrams for RHS under combined bending and transverse bearing force

### *6.3.6 Examples*

*6.3.6.1 Example 1*

A cold-formed 100 × 50 × 2.5 RHS section is subjected to an interior design bearing force of 10 kN. The force is applied over the full width of the RHS and for a length of 50 mm along the RHS. Assume that the yield stress is 355 N/mm$^2$. Check the bearing capacity of the beam.

**Solution according to AS 4100**

1. Dimensions and Properties:

$d = 100$ mm
$b = 50$ mm
$t = 2.5$ mm
$b_s = 50$ mm

From Equation (6.1b) the external corner radius of the chord
$r_{ext} = 2\,t = 2 \times 2.5 = 5$ mm

From Equation (6.7a) the dimension $b_b$ can be determined as
$b_b = b_s + d + 3 \cdot r_{ext} = 50 + 100 + 3 \times 5 = 165$ mm

$f_y = 355$ N/mm$^2$

2. Web bearing buckling versus web bearing yield

The critical $\left(\dfrac{d - 2 \cdot r_{ext}}{t}\right)$ for interior bearing with $t < 3$ mm can be determined from Equation (6.13a) or from Table 6.1 as 32.8 for $f_y = 355$ N/mm$^2$.

$$\text{Since } \left(\frac{d - 2 \cdot r_{ext}}{t}\right) = \left(\frac{100 - 2 \times 5}{2.5}\right) = 36 > \left(\frac{d - 2 \cdot r_{ext}}{t}\right)_{critical} = 32.8$$

web bearing buckling governs, i.e.

$R_b = R_{bb}$

3. Web bearing capacity

$$R_b = R_{bb} = 2 \cdot b_b \cdot t \cdot f_y \cdot \alpha_c$$

$\alpha_c$ can be determined from Figure 6.12 using the $\lambda_n$ defined in Equation (6.8a):

$$\lambda_n = 3.5 \cdot \left(\frac{d - 2 \cdot r_{ext}}{t}\right) \cdot \sqrt{\frac{f_y}{250}} = 3.5 \times 36 \cdot \sqrt{\frac{355}{250}} = 150$$

Hence $\alpha_c \approx 0.28$

$$R_b = R_{bb} = 2 \cdot b_b \cdot t \cdot f_y \cdot \alpha_c = 2 \times 165 \times 2.5 \times 355 \times 0.28 = 82{,}005 \text{ N} = 82.0 \text{ kN}$$

$$\phi R_b = 0.9 \times 82 = 73.8 \text{ kN} > R^* = 10 \text{ kN}$$

The cold-formed $100 \times 50 \times 2.5$ RHS is satisfactory.

**Note**: For the interest of readers the web bearing yield capacity ($R_{by}$) is calculated using Equation (6.10) as follows.

$$k_v = \frac{d - 2 \cdot r_{ext}}{t} = \frac{100 - 2 \times 5}{2.5} = 36$$

$$k_s = \frac{2 \cdot r_{ext}}{t} - 1 = \frac{2 \times 5}{2.5} - 1 = 3$$

$$\alpha_{pm} = \frac{1}{k_s} + \frac{0.5}{k_v} = \frac{1}{3} + \frac{0.5}{36} = 0.347$$

$$1 - \alpha_{pm}^2 = 1 - 0.347^2 = 0.880$$

$$\frac{0.5}{k_s} = \frac{0.5}{3} = 0.167$$

$$\frac{k_s}{k_v} = \frac{3}{36} = 0.0833$$

$$\frac{0.25}{k_v^2} = \frac{0.25}{36^2} = 0.0001929$$

$$\alpha_p = \frac{0.5}{k_s} \cdot \left[ 1 + (1 - \alpha_{pm}^2) \cdot \left( 1 + \frac{k_s}{k_v} - (1 - \alpha_{pm}^2) \cdot \frac{0.25}{k_v^2} \right) \right]$$
$$= 0.167 \cdot [1 + 0.880 \cdot (1 + 0.0833 - 0.880 \cdot 0.0001929)]$$
$$= 0.326$$

$$R_{by} = 2 \cdot b_b \cdot t \cdot f_y \cdot \alpha_p = 2 \times 165 \times 2.5 \times 355 \times 0.326 = 95{,}477 \text{ N} = 97.5 \text{ kN}$$

It can be seen that $R_{by}$ is larger than $R_{bb}$ of 82 kN.

Therefore the prediction of web bearing buckling using critical $\left( \frac{d - 2 \cdot r_{ext}}{t} \right)$ is correct.

**Solution according to BS 5950 Part 1**

1. Dimensions and Properties:

$d$ = 100 mm
$b$ = 50 mm
$t$ = 2.5 mm
$b_s$ = 50 mm
$f_y$ = 355 N/mm$^2$

2. Web bearing buckling capacity

$d/b$ = 100/50 = 2

From Equation (6.9d)

$$e = 0.026 \cdot b + 0.978 \cdot t + 0.002 \cdot d = 0.026 \times 50 + 0.978 \times 2.5 + 0.002 \times 100 = 3.945 \text{ mm}$$

From Equation (6.9c)

$$K = \left[\frac{3\cdot(3+\frac{d}{b})}{(\frac{d}{b})^2+4\cdot(\frac{d}{b})+3}\right]\cdot\left(\frac{e}{t}\right)\cdot\left(\frac{b-t-e}{b-t}\right)$$

$$= \left[\frac{3\cdot(3+2)}{2^2+4\cdot2+3}\right]\cdot\left(\frac{3.945}{2.5}\right)\cdot\left(\frac{50-2.5-3.945}{50-2.5}\right)$$

$$= 1.0\times1.578\times0.917$$

$$= 1.447$$

web slenderness $\lambda = 0.75\cdot\sqrt{12}\cdot\left(\frac{d-2\cdot t}{t}\right) = 0.75\cdot\sqrt{12}\cdot\left(\frac{100-2\times2.5}{2.5}\right) = 98.7$

$p_c$ can be calculated using Equation (4.15) or Figure 4.9 for yield stress of 355 N/mm$^2$, i.e.

$p_c \approx 172\,\text{N/mm}^2$

Hence $\frac{p_c}{f_y} = \frac{172}{355} = 0.485$

From Equation (6.9b)

$$\alpha_{SCI} = \frac{\frac{p_c}{f_y}}{1+K\cdot\frac{p_c}{f_y}} = \frac{0.485}{1+1.447\times0.485} = 0.285$$

From Equation (6.9a)

$R_{bb} = 2\cdot(b_s+d)\cdot t\cdot f_y\cdot\alpha_{SCI} = 2\cdot(50+100)\cdot2.5\times355\times0.285 = 75{,}881\text{ N} = 75.9\text{ kN}$

This value of 75.9 kN is comparable with that of 82.0 kN predicted by AS 4100.

3. Web bearing yield capacity

From Equation (6.11a)

$R_{by} = 2\cdot(b_s+5\cdot t)\cdot t\cdot f_y = 2\cdot(50+5\cdot2.5)\cdot2.5\times355 = 110{,}938\text{ N} = 111\text{ kN}$

This value of 111 kN is about 14% larger than that of 97.5 kN predicted by AS 4100. The lower value given by AS 4100 may be due to the fact that solution of AS 4100 is based a plastic mechanism analysis where the external corner radius is taken into account.

4. Web bearing capacity

From Equation (6.12)

$$R_b = \min\{R_{bb}, R_{by}\} = R_{bb} = 75.9\ kN$$

It is noted that the web bearing buckling governs for both AS 4100 and BS 5950 Part 1.

*6.3.6.2 Example 2*

The simply supported beam shown Figure 6.17 has two concentrated loads ($R^* = 10$ kN) applied in the same way as described in Section 6.3.6.1 Examples 1. Full lateral restraint is applied at the location of the loads. The beam size is cold-formed RHS 100×50×2.5. Assume that the yield stress is 355 N/mm$^2$.

(i) Check the beam is adequate when the distance between the applied $R^*$ and the end support is 0.5 m.

(ii) If the two loads ($R^*$) move towards each other, what is the maximum distance ($L$) between the applied load $R^*$ and the end support before failure occurs under combined bending and bearing?

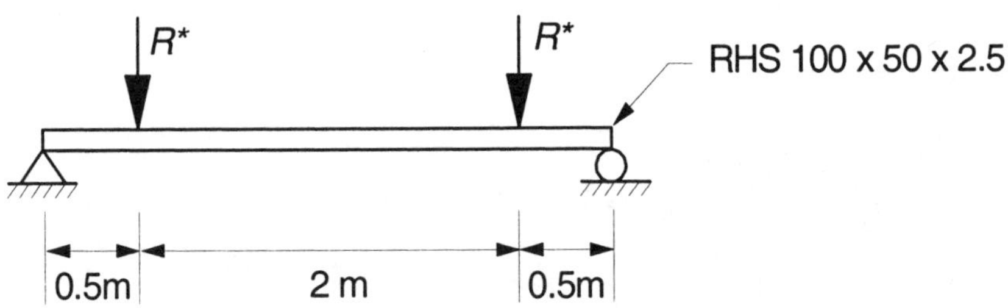


Figure 6.17 A beam under two concentrated forces applied through bearing plates

**Solution according to AS 4100**

(i) The interaction between the bending and concentrated force should be checked against Equation (6.14).

1. Dimensions and Properties:

$d = 100$ mm
$b = 50$ mm
$t = 2.5$ mm
$b_s = 50$ mm
$f_y = 355$ N/mm$^2$

2. Calculate the two ratios for selecting the appropriate interaction formula

$$\frac{b_s}{b} = \frac{50}{50} = 1.0$$

$$\frac{d - 2 \cdot t}{t} = \frac{100 - 2 \times 2.5}{2.5} = 38 > 30$$

Therefore Equation (6.14b) should be used, i.e.

$$0.8 \cdot \left( \frac{R^*}{\phi \cdot R_b} \right) + \left( \frac{M^*}{\phi \cdot M_s} \right) \leq 1.0$$

3. Calculate design forces and resistances

$R^* = 10$ kN
$\phi R_b = 73.8$ kN (from Section 6.3.6.1 Example 1)

$M^* = R^* \times 0.5 = 10 \text{ kN} \times 0.5 \text{ m} = 5 \text{ kNm}$
$\phi M_s = \phi Z_e f_y$

$\phi = 0.9$ (from Table 3.4 of AS 4100)

For RHS 100 × 50 × 2.5 the effective section modulus $Z_{ex} = 22.7 \times 10^3 \text{ mm}^3$ (from ASI 1999)

$\phi M_s = 0.9 \times 22.7 \times 10^3 \times 355 = 7.25 \times 10^6 \text{ Nmm} = 7.25 \text{ kNm}$

4. Check interaction:

$$0.8 \cdot \left( \frac{R^*}{\phi \cdot R_b} \right) + \left( \frac{M^*}{\phi \cdot M_s} \right) = 0.8 \cdot \left( \frac{10}{73.8} \right) + \left( \frac{5}{7.25} \right) = 0.80 < 1.0$$

Therefore the cold-formed RHS 100 × 50 × 2.5 is adequate.

(ii) Similar to Part (i)

1. Dimensions and Properties:

$d = 100$ mm
$b = 50$ mm
$t = 2.5$ mm
$b_s = 50$ mm
$f_y = 355 \text{ N/mm}^2$

2. Calculate the two ratios for selecting the appropriate interaction formula

$$\frac{b_s}{b} = \frac{50}{50} = 1.0$$

$$\frac{d - 2 \cdot t}{t} = \frac{100 - 2 \times 2.5}{2.5} = 38 > 30$$

Therefore Equation (6.14b) should be used, i.e.

$$0.8 \cdot \left( \frac{R^*}{\phi \cdot R_b} \right) + \left( \frac{M^*}{\phi \cdot M_s} \right) \leq 1.0$$

3. Calculate design forces and resistances

$R^* = 10$ kN
$\phi R_b = 73.8$ kN (from Section 6.3.6.1 Example 1)

$M^* = R^* \times \mathrm{L} = 10\ \text{kN} \times L\ \text{m} = 10L\ \text{kNm}$
$\phi M_s = 7.25$ kNm

4. Determine the maximum distance ($L$)

From interaction formula:

$$0.8 \cdot \left( \frac{R^*}{\phi \cdot R_b} \right) + \left( \frac{M^*}{\phi \cdot M_s} \right) = 0.8 \cdot \left( \frac{10}{73.8} \right) + \left( \frac{10 \cdot L}{7.25} \right) \leq 1.0$$

$$0.108 + 1.379 \cdot L \leq 1.0$$

$$L \leq \frac{1.0 - 0.108}{1.379} = 0.65\ \text{m} = 650\ \text{mm}$$

Therefore the maximum distance ($L$) is 650 mm.

## 6.4 References

1. ASI (1999), Design Capacity Tables for Structural Steel – Volume 2: Hollow Sections, Australian Steel Institute, Sydney, Australia
2. BSI (1998), Structural use of Steelwork in Building, BS 5950, Part 5, British Standard Institution, London, UK
3. BSI (2000), Structural use of Steelwork in Building, BS 5950, Part 1, British Standard Institution, London, UK
4. CSA-S136 (2001), Cold-Formed Steel Structural Members, CSA-S136-01, Canadian Standards Association, Toronto, Ontario, Canada

5. Hancock, G.J., Sully, R.M. and Zhao, X.L. (1994), Hollow Flange Beams and RHS Sections under Combined Bending and Bearing, In: *Tubular Structures VI*, Grundy, P., Holgate, A. and Wong B. (eds), Balkema: Rotterdam, The Netherlands, pp 47-54
6. Hetrakul, N. and Yu, W.W. (1978), Structural Behaviour of Beam Webs Subjected to Web Crippling and a Combination of Web Crippling and Bending, *Final Report*, Civil Engineering Study 78-4, June, University of Missouri-Rolla, Missouri, USA
7. Hetrakul, N. and Yu, W.W. (1979), Cold-Formed Steel I-Beams Subjected to Combined Bending and Web Crippling, In: *Thin-Walled Structures – Recent Technical Advances and Trends in Design, Research and Construction*, Rhodes, J. and Walker, A.C. (eds), Granada Publishing: London, UK, pp 413-426
8. Kato, B. and Nishiyama, I. (1979), The Static Strength of R.R.-Joints with Large b/B ratio, *CIDECT Report 5Y,* Department of Architecture, Faculty of Engineering, University of Tokyo, Tokyo, Japan
9. Kato, B. and Nishiyama, I. (1980), T-Joints Made of Rectangular Tubes, *Proceedings*, 5th International Specialty Conference on Cold-Formed Steel Structures, St. Louis, Missouri, USA, pp 663-679
10. NAS (2002), North American Specification for the Design of Cold-Formed Steel Structural Members, American Iron and Steel Institute, Washington D.C, USA
11. Packer, J.A. (1984), Web Crippling of Rectangular Hollow Sections, *Journal of Structural Engineering,* ASCE, 110(10), pp 2357-2373
12. Packer, J.A., Wardenier, J., Kurobane, Y., Dutta, D. and Yeomans, N. (1992), *Design Guide for Rectangular Hollow Section (RHS) Joints under Predominantly Static Loading*, TÜV-Verlag, Köln, Germany
13. Ratcliff, G.D. (1975), Interaction of Concentrated Loads and Buckling in C-Shaped Beams, *Proceedings*, Third International Specialty Conference on Cold-Formed Steel Structures, November, University of Missouri-Rolla, Missouri, USA, pp 337-356
14. Roberts, T.M. (1981), Slender Plate Girders Subjected to Edge Loading, *Proceedings,* Institution of Civil Engineers, Volume 71, Part 2, September, pp 805-809
15. Roberts, T.M. (1983), Patch Loading on Plate Girders, Chapter 3 in *Plated Structures: Stability and Strength,* R. Narayanan (ed), Applied Science Publishers: London, UK, pp 77-102
16. Rockey, K.C., El-Gaaly, M.A. and Bagchi, D.K. (1972), Failure of Thin-Walled Members under Patch Loading, *Journal of the Structural Division,* ASCE, 98(ST12), pp 2739-2752
17. SCI (2002), Steelwork Design Guide to BS 5950-1:2000, Volume 1, Section Properties and Members Capacities, The Steel Construction Institute, UK
18. Standards Australia (1996), Cold-Formed Steel Structures, Australian Standard AS/NZS 4600, Standards Australia, Sydney, Australia
19. Standards Australia (1998), Steel Structures, Australian Standard AS 4100, Standards Australia, Sydney, Australia
20. Trahair, N.S., Bradford, M.A. and Nethercot, D.A. (2001), *The Behaviour and Design of Steel Structures to BS 5950*, Third Edition – British, Spon Press, London, UK

21. Wing, B.A. and Schuster, R.M. (1986), Web Crippling of Multi-Web Deck Sections Subjected to Interior One Flange Loading, *Proceedings*, 8th International Specialty Conference on Cold-Formed Steel Structures, St. Louis, Missouri, USA, pp 371-402
22. Zhang, Z.L., Shen, Z.Y. and Chen, X.C. (1989), Nonlinear FEM Analysis and Experimental Study of Ultimate Capacity of Welded RHS Joints, *Proceedings*, 3rd International Symposium on Tubular Structures, Lappeenranta, Finland, Session P4.03-1, pp 1-9
23. Zhao, X.L. (1992), The Behaviour of Cold-Formed Rectangular Hollow Section Beams under Combined Actions, *PhD Thesis*, The University of Sydney, Sydney, Australia
24. Zhao, X.L. (2000), Deformation Limit and Ultimate Strength of Welded T-joints in Cold-Formed RHS Sections, *Journal of Constructional Steel Research*, 53(2), pp 149-165
25. Zhao, X.L. and Hancock, G.J. (1991a), T-Joints in Rectangular Hollow Sections Subject to Combined Actions, *Journal of Structural Engineering,* ASCE, 117(8), pp 2258-2277
26. Zhao, X.L. and Hancock, G.J. (1991b), Plastic Mechanism Analysis of T-Joints in RHS Under Concentrated Force, *Journal of Singapore Structural Steel Society,* 2(1), pp 31-44
27. Zhao, X.L. and Hancock, G.J. (1992a), Square and Rectangular Hollow Sections Subject to Combined Actions, *Journal Structural Engineering*, ASCE, 118(3), pp 648-668
28. Zhao, X.L. and Hancock, G.J. (1992b), Design Formulae for Web Crippling of Rectangular Hollow Sections, *Proceedings*, 3rd Pacific Structural Steel Conference, Tokyo, Japan, pp 97-104
29. Zhao, X.L. and Hancock, G.J. (1995), Square and Rectangular Hollow Sections under Transverse End Bearing Force, *Journal of Structural Engineering*, ASCE, 121(11), pp 1565-1573
30. Zhao, X.L., Hancock, G.J. and Sully R.M. (1996), Design of Tubular Members and Connections using Amendment No.3 to AS 4100, *Steel Construction*, Journal of the Australian Institute of Steel Construction, 30(4), pp 2-15

# Chapter 7: Tension Members and Welds in Thin Cold-Formed Tubes

## 7.1 Tension Members

Tension members are those structural elements which are subjected to direct axial loads that tend to elongate the members. They occur as components of trusses, hangers and cables for floors or roofs, in bracing systems, as tie rods, and similar members. The design of tension members is relatively simpler than the design of members in bending or compression. In limit states design, the basic requirement is simply that enough cross-sectional area be provided in order that the factored resistance of the member is equal to or greater than the factored load in the member.

Most tubular tension members contain connections which are welded so that the connection strength may limit the tension member strength. For this reason, simple welded connections are included in this chapter.

### *7.1.1 AS 4100 (Standards Australia 1998)*

The nominal section capacity of a tubular member in tension shall be taken as the lesser of

$$N_t = A_g \cdot f_y \tag{7.1a}$$

and

$$N_t = A_n \cdot 0.85 \cdot f_u \tag{7.1b}$$

where $A_g$ is the gross area of the tube, $f_y$ is the yield stress of the tube, $A_n$ is the net area of the cross-section which accounts for deductions for bolt holes at specific sections and $f_u$ is the ultimate tensile strength of the tube.

It can be seen that yielding of the gross area and fracture through the net area constitute strength limit states. The 0.85 factor in the expression for net section fracture is intended to account for sudden failure by local brittle behaviour at the net section.

A capacity factor ($\phi$) of 0.90 is adopted in AS 4100 for designing tension members.

### *7.1.2 BS 5950 Part 1 (BSI 2000)*

A single equation is used in BS 5950 Part 1 to represent the two separate limit states of the yield of the gross section and fracture of the net section The tension capacity of a tube can be calculated as

$$N_t = p_y \cdot A_e \tag{7.2}$$

where $p_y$ is the design yield stress and $A_e$ is the effective net area.

The design yield stress shall be taken as $f_y$ but should not be greater than $f_u/1.2$.

The condition of $p_y \leq f_u /1.2 = 0.83f_u$ was used in BS 5950 Part 1 instead of checking both yielding and fracture failure as in AS 4100. It seems that the tension capacity given in BS 5950 Part 1 is smaller than that given in AS 4100 when $f_y$ is between $0.83f_u$ and $0.85f_u$. Also, the yield condition in BS 5950 Part 1 applies to the effective area rather than the gross area as in AS 4100.

## 7.2 Characteristics of Welds in Thin Cold-Formed Tubes

Gas metal arc welding (GMAW also called MIG welding) and tungsten-arc welding (TIG welding) are two commonly used methods for joining thin-walled sections. The material properties (strength, ductility and hardness) may change in the heat-affected-zone (HAZ) due to the welding. It seems that when the yield stress of the tubes is less than 460 N/mm$^2$ the influence of welding on the tube properties in the HAZ is not significant (Zhao and Jiao 2004). All welding procedures should be prequalified according to certain standards, e.g. AS/NZS 1554.1 (Standards Australia 2000) in Australia. Two categories of welds are specified in AS 4100 for design, namely SP (structural purpose) weld and GP (general purpose) weld. A more stringent requirement is needed for SP welds when conducting macro cross-section examination tests according to AS 2205.5.1 (Standards Australia 2003a). A larger capacity factor is specified in AS 4100 for SP weld design than for GP weld design.

The previous version of the Australian Standard for Steel Structures AS 4100 (Standards Australia 1990) only applied to steel that is 3 mm or thicker. Tubular sections less than 3 mm thick could not be designed to that standard, and had to be designed to the Standard for Cold-Formed Steel Structures AS 1538-1988 (Standards Australia 1988). As described in Chapters 3, 4, 5 and 6 member design rules have been developed and included in AS 4100 (Standards Australia 1998) for tubular sections less than 3 mm as summarised in Zhao and Hancock (1998). However, the design rules for welded connections in AS 1538-1988 for members less than 3 mm thick may be inappropriate. The design rules in AS 1538-1988 were the same as those in the American Iron and Steel Institute Specification (1986), which were based on tests performed by Peköz and McGuire (1981) on sheet steel. Similar research on welded sheet steel was also performed by Stark and Soetens (1980), which was adopted in Eurocode 3 Part 1.3 (2004). Because they were derived for flat sheet steel, these design rules may be inappropriate when applied to RHS. The thickness limits for the design of welded connections in various standards vary from 3 mm to 4.6 mm, as summarised in Table 7.1. In general there are two sets of standards. One is for sections greater than the thickness limit while the other is for sections less than the thickness limit.

Table 7.1 Summary of thickness limit in designing welded connections

| Country/Region | Standard | Thickness limit |
|---|---|---|
| America | AISC specification (1999) | ≥ 4.6 mm (or 3/16 inch) |
| | NAS (2004) | < 4.6 mm (or 3/16 inch) |
| | AWS D1.1 (2002) | ≥ 3.2 mm (or 1/8 inch) |
| | AWS D1.3 (1998) | < 3.2 mm (or 1/8 inch) |
| Australia | AS 4100-1990 | ≥ 3 mm |
| | AS 4100-1998 | ≥ 3 mm except cold-formed tubes for which all thicknesses apply |
| | AS 1538-1988<br>AS/NZS 4600-1996 | < 3 mm<br>< 3 mm |
| Canada | CSA-S16-01 (2001) | ≥ 3.5 mm |
| | CSA-S136-01 (2001) | < 3.5 mm |
| Europe | Eurocode 3 Part 1.1 (2003) | ≥ 4.0 mm |
| | Eurocode 3 Part 1.3 (2004) | < 4.0 mm |

Research was conducted on welded connections in thin-walled cold-formed RHS by Zhao and Hancock (1995a, 1995b, 1996) and Zhao et al (1999). The intention was to include design rules for welded connections less than 3 mm into AS 4100-1998 (Zhao et al 1996). Three types of welds were investigated namely butt welds, transverse fillet welds and longitudinal fillet welds, as shown in Figure 7.1. The nominal yield stress of RHS used in the test program was 350 $N/mm^2$ and 450 $N/mm^2$. All the welding procedures were prequalified according to Section 4 of the Australian Standard AS 1554.1-1991. For butt welds only those with complete (or full) penetration were studied. For fillet welds the actual weld size (leg length) was found to be about 1.5 times the nominal weld size. This matches the trend identified by Pham and Bennetts (1983) after analyzing the data obtained from the international test series with participation from 10 countries (Ligtenberg 1968), as shown in Figure 7.2. The oversized welding was used in the calibration of design formulae for fillet welds. Design rules for welds in thin sections were calibrated using the reliability analysis technique used to calibrate AS 4100-1998 for thicker sections. As shown later in this chapter, the design rule in BS 5950 Part 1 is similar to that in AS 4100 for complete (or full) penetration butt welds design, i.e. checking the parent metal strength. The design rules in BS 5950 Part 1 are also similar to those in AS 4100 in designing fillet welds, i.e. checking the weld metal strength.

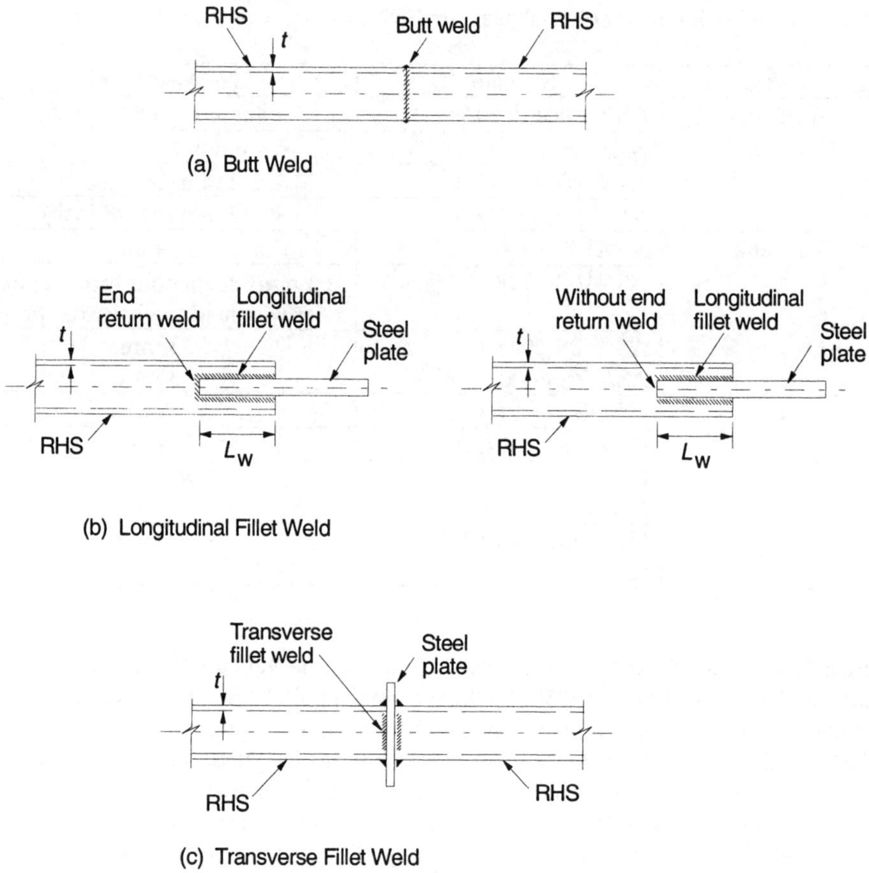


Figure 7.1 Three types of welds for design: (a) Butt welds, (b) Longitudinal fillet welds, (c) Transverse fillet welds

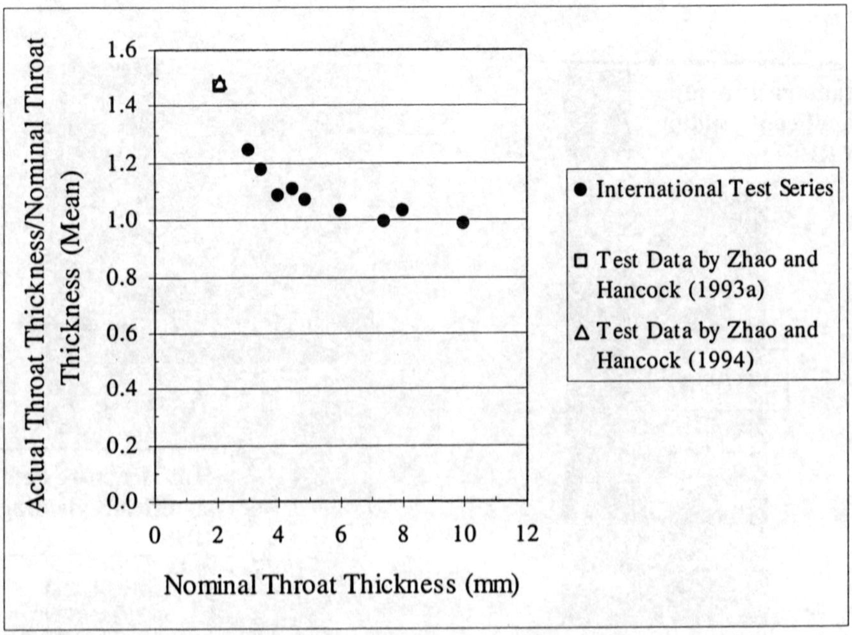


Figure 7.2 Variation in throat thickness

## 7.3 Butt Welds

### *7.3.1 Fracture After or Before Significant Yielding*

Two different failure modes were observed for butt welded RHS members with a nominal yield stress of 350 N/mm$^2$ (Zhao and Hancock 1993a) and 450 N/mm$^2$ (Zhao and Hancock 1994). One is seam fracture after significant yielding of the RHS as shown in Figure 7.3 (a). The other is HAZ fracture before significant yielding of the RHS as illustrated in Figure 7.3 (b). It is interesting to note that the tearing in Figure 7.3 (a) occurs along the seam of the RHS section. This may be due to the lower ductility of the seam. No significant necking was observed in Figure 7.3 (b) where a crack initiated at the corners of the RHS. This may be due to the lower ductility of RHS corners. The failure mode in Figure 7.3 (a) is governed by Equation (7.1a) whereas the failure mode in Figure 7.3 (b) is governed by Equation (7.1b). In BS 5950 Part 1, the failure mode in Figure 7.3 (b) is governed by Equation (7.2) with a limiting design yield stress not greater than $f_u/1.2 = 0.83f_u$.

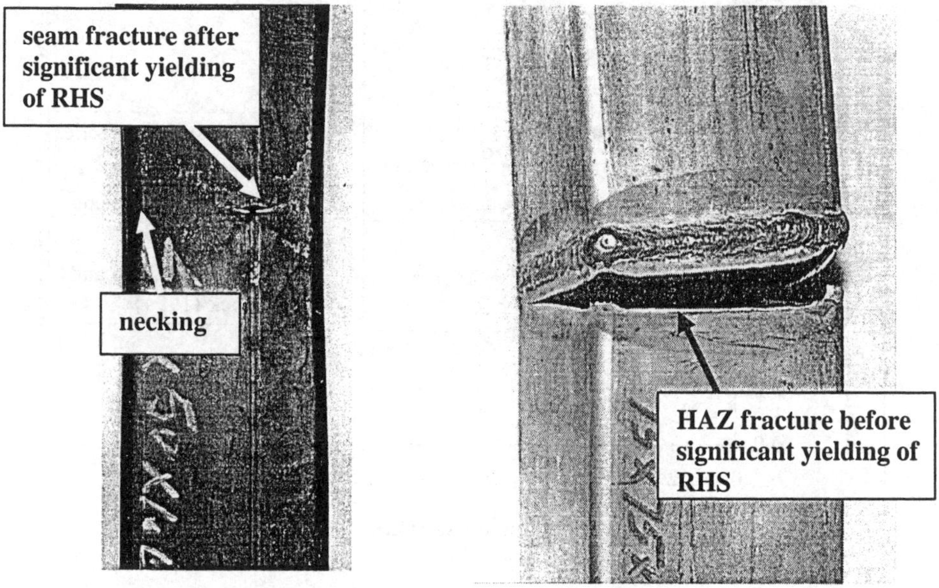


(a) Butt welded RHS with a nominal yield stress of 350 $N/mm^2$

(b) Butt welded RHS with a nominal yield stress of 450 $N/mm^2$

Figure 7.3 Failure modes (fracture after and before significant yielding of RHS)

### *7.3.2 Design Rules*

#### *7.3.2.1 AS 4100*

Reliability analyses were conducted (Zhao and Hancock 1995a, 1994) to calibrate the design rules for butt welded RHS members. It was concluded that the complete penetration butt welds in RHS can be designed according to the tension member rules in AS 4100-1998 with a capacity factor ($\phi$) of 0.90, i.e. Equations (7.1a) and (7.1b) described in Section 7.1.

#### *7.3.2.2 BS 5950 Part 1*

Similarly the tension member rules (Equation (7.2)) can be applied to tubular members joined by complete penetration butt-welds.

It should be pointed out that although the connection strength of a welded connection is based on the parent metal strength, the weld metal strength is required to match the parent metal strength. Generally, matching requires the strength of the weld metal to be equal to or slightly larger than that of the parent metal (Funderburk 2001, Kyuba et al 2001). In both AS 4100 and BS 5950 Part 1, a matching electrode is required.

### 7.3.3 *Examples*

#### *7.3.3.1 Example 1*

A tension member with a full perimeter welded connection is subjected to an axial tension force of 260 kN. Check the suitability of using a 75 × 75 × 2.5 RHS Grade C350 with a complete penetration butt weld.

**Solution according to AS 4100**

1. Dimensions and Properties

$d = 75$ mm
$b = 75$ mm
$t = 2.5$ mm

$A_g = A_n = 709$ mm$^2$ based on actual rounded corners
$f_y = 350$ N/mm$^2$
$f_u = 430$ N/mm$^2$ (from Table 2.4 of Chapter 2)

2. Nominal tension capacity

Since $f_y = 350$ N/mm$^2$ $< 0.85 f_u = 0.85 \times 430 = 366$ N/mm$^2$, then the nominal tension capacity is given by Equation (7.1a), i.e.

$$N_t = A_g \cdot f_y = 709 \times 350 = 248{,}150 \text{ N} = 248 \text{ kN}$$

3. Design tension capacity

$$\phi N_t = 0.9 \times 248 = 223 \text{ kN}$$

4. Design check

Design force $N^* = 260$ kN

Since $\phi N_t = 233 \text{ kN} < N^* = 260 \text{ kN}$, the 75 × 75 × 2.5 RHS – Grade C350 with a complete penetration butt weld is not satisfactory.

**Solution according to BS 5950 Part 1**

1. Dimensions and Properties

$d = 75$ mm
$b = 75$ mm
$t = 2.5$ mm

$A_e = 709$ mm$^2$ based on actual rounded corners
$f_y = 350$ N/mm$^2$
$f_u = 430$ N/mm$^2$ (from Table 2.4 of Chapter 2)

2. Design yield stress

Since $f_y = 350\ \text{N/mm}^2 < f_u/1.2 = 430/1.2 = 358\ \text{N/mm}^2$

$p_y = f_y = 350\ \text{N/mm}^2$

3. Tension capacity

From Equation (7.2)

$$N_t = p_y \cdot A_e = 350 \times 709 = 248{,}150\ \text{N} = 248\ \text{kN}$$

4. Design check

Design force $N^* = 260$ kN

Since $N_t = 248\ \text{kN} < N^* = 260\ \text{kN}$, the 75 × 75 × 2.5 RHS – Grade C350 with a complete penetration butt weld is not satisfactory.

Note: In this example, the nominal tension capacity given in AS 4100 is the same as the tension capacity given in BS 5950 Part 1. Both calculations show that the 75 × 75 × 2.5 RHS – Grade C450 with a complete penetration butt weld is not satisfactory.

*7.3.3.2 Example 2*

A tension member with a full perimeter welded connection is subjected to an axial tension force of 260 kN. Check the suitability of using a 75 × 75 × 2.5 RHS Grade C450 with a complete penetration butt weld.

**Solution according to AS 4100**

1. Dimensions and Properties

$d = 75$ mm
$b = 75$ mm
$t = 2.5$ mm

$A_g = A_n = 709\ \text{mm}^2$ based on actual rounded corners
$f_y = 450\ \text{N/mm}^2$
$f_u = 500\ \text{N/mm}^2$ (from Table 2.4 of Chapter 2)

2. Nominal tension capacity

Since $f_y = 450\ \text{N/mm}^2 > 0.85 f_u = 0.85 \times 500 = 425\ \text{N/mm}^2$, then the nominal tension capacity is given by Equation (7.1b), i.e.

$$N_t = A_n \cdot 0.85 f_u = 709 \times 0.85 \times 500 = 301{,}325\ \text{N} = 301\ \text{kN}$$

3. Design tension capacity

$\phi N_t = 0.9 \times 301 = 271$ kN

4. Design check

Design force $N^* = 260$ kN

Since $\phi N_t = 271$ kN $> N^* = 260$ kN, the 75 × 75 × 2.5 RHS – Grade C450 with a complete penetration butt weld is satisfactory.

**Solution according to BS 5950 Part 1**

1. Dimensions and Properties

$d = 75$ mm
$b = 75$ mm
$t = 2.5$ mm

$A_e = 709$ mm$^2$ based on actual rounded corners
$f_y = 450$ N/mm$^2$
$f_u = 500$ N/mm$^2$ (from Table 2.4 of Chapter 2)

2. Design yield stress

Since $f_y = 450$ N/mm$^2 > f_u/1.2 = 500/1.2 = 417$ N/mm$^2$

$p_y = 417$ N/mm$^2$

3. Tension capacity

From Equation (7.2)

$$N_t = p_y \cdot A_e = 417 \times 709 = 295{,}653 \text{ N} = 296 \text{ kN}$$

4. Design check

Design force $N^* = 260$ kN

Since $N_t = 296$ kN $> N^* = 260$ kN the 75 × 75 × 2.5 RHS – Grade C450 with a complete penetration butt weld is satisfactory.

Note: In this example, the nominal tension capacity (301 kN) given in AS 4100 is slightly higher than the tension capacity (296 kN) given in BS 5950 Part 1. This is because $0.85f_u$ is used in AS 4100 whereas $0.83f_u$ ($=f_u/1.2$) is used in BS 5950 Part 1 for design stress in the case of fracture failure. Both calculations show that the 75 × 75 × 2.5 RHS – Grade C450 with a complete penetration butt weld is satisfactory.

## 7.4 Longitudinal Fillet Welds

### *7.4.1 Failure Modes*

Three types of failure may occur in slotted gusset plate connections to steel RHS and CHS (circular hollow section), namely block shear tear-out (TO) failure of steel tubes along the weld, shear lag (SL) failure causing tubes to fail circumferentially and section failure without any shear lag reduction. A typical TO failure mode and SL failure mode in welded CHS are shown in Figure 7.4. Section failure without any shear lag reduction can be seen in Figure 7.3 (a).

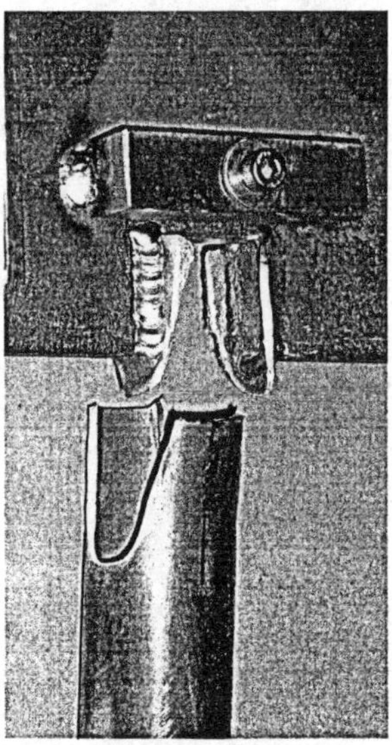

(a) TO failure

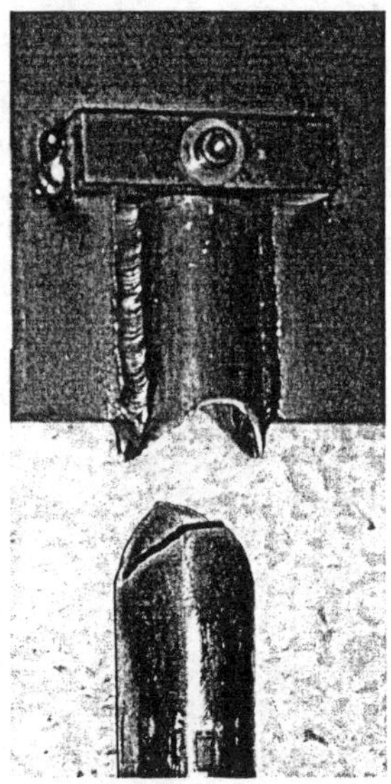

(b) SL failure

Figure 7.4 Typical block tear-out (TO) failure mode and shear lag (SL) failure mode (Ling et al 2004)

In AS 4100 and BS 5950 Part 1 the design of longitudinal fillet welds and the design of transverse fillet welds are all based on the failure of weld metal. No shear lag reduction factor is given. However the commentary to Clause 9.7.3.10 of AS 4100 and Clause 6.7.2.4 of BS 5950 Part 1 both require that when longitudinal fillet welds are used alone in a connection $L_w$ should be at least equal to the width of the connecting material because of shear lag. This applies only to lap joints connecting steel plates. No information is given specifically for slotted gusset plate connections to RHS or CHS.

Recent test data in Korol (1996) showed that no shear lag reduction factor is needed for slotted gusset connections to RHS when $L_w$ is larger than $1.2w$ where $w$ is defined in Figures 7.5 (a) and 7.5 (b) and can be expressed approximately as:

$$w = d + b - T \tag{7.3}$$

where $d$ is the overall depth of the RHS, $b$ is the overall width of the RHS and $T$ is the thickness of the plate.

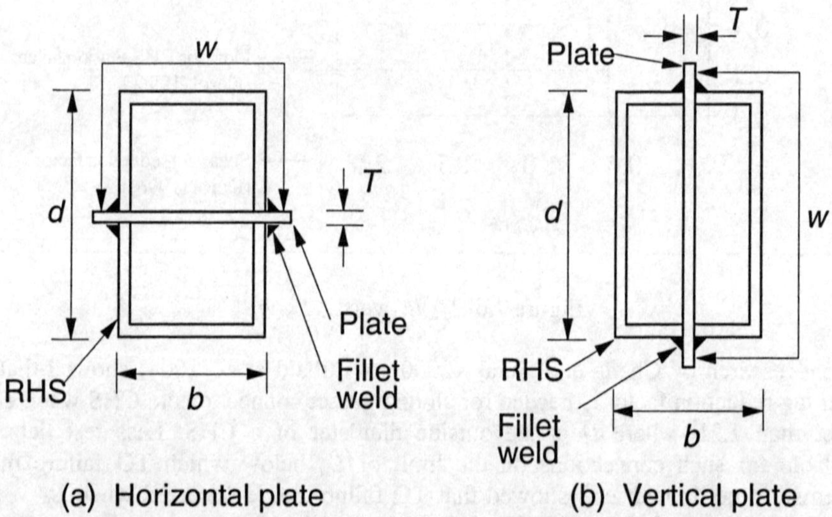


Figure 7.5 Definition of $w$ – distance between the welds measured around the perimeter of the tube

It was also shown in Korol (1996) that tear-out (TO) failure mode governs when $L_w$ is less than $0.6w$. Both conclusions are confirmed by the test data in Zhao and Hancock (1993a, 1993b, 1994), Zhao et al (1999) for thin RHS ($t \leq 3$ mm), as shown in Figure 7.6 where $P_{max}$ is the maximum load carrying capacity obtained in the test and $N_t$ is defined in Equation (7.1) with measured dimensions and material properties used in the calculations.

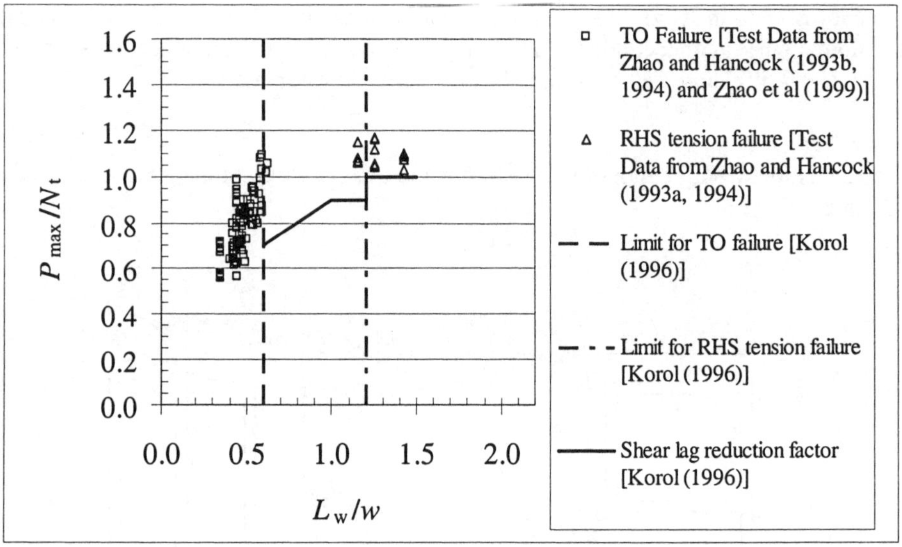


Figure 7.6 $P_{max}/N_t$ versus $L_w/w$

Recent research by Cheng and Kulak (2000), Willibald et al (2004) showed that no shear lag reduction factor is needed for slotted gusset connections to CHS when $L_w$ is larger than $1.3D$ where $D$ is the outside diameter of a CHS. Less test data are available for such connections on the limit of $L_w$ below which TO failure mode governs. Some limited tests showed that TO failure mode occurred when $L_w$ is less than $0.7w$ (Willibald et al 2004) or $0.6w$ (Ling et al 2004) where $w$ is the distance between welds measured around the perimeter of the CHS, defined in a similar way as that in Figure 7.5.

The condition to determine failure modes for slotted gusset-plate connections to cold-formed tubes is summarised in Table 7.2.

Table 7.2 Failure modes for slotted gusset-plate connections to cold-formed tubes

| Failure mode | Condition |
|---|---|
| Tear-out failure | $L_w < 0.6w$ for both RHS and CHS |
| Shear lag failure | $0.6 \cdot w \le L_w \le 1.2 \cdot w$ for RHS<br>$0.6 \cdot w \le L_w \le 1.3 \cdot D$ for RHS |
| RHS tension failure without shear lag reduction | $L_w > 1.2w$ for RHS<br>$L_w > 1.3D$ for CHS |

Various formulae were proposed in the literatures for block shear tear-out failure and shear lag reduction factors (Packer and Henderson 1997). It is outside the scope of this book to completely cover this topic. More references are given in Section 9.5. Only the formulae given by Korol (1996) for slotted gusset plate connections to RHS are given here as an example:

For $L_w < 0.6w$, parent metal shear resistance $V_r$ will govern, provided that the fillet weld size is nongoverning. In this case,

$$V_r = \phi \cdot 0.67 \cdot f_y \cdot L \cdot t \quad (7.4)$$

where $f_y$ is the yield stress of the RHS, $L$ is the total weld length, $t$ is the tube wall thickness and $\phi$ is the capacity factor taken as 0.9.

For $L_w \geq 0.6w$, shear lag resistance governs and is given by

$$T_r = \phi \cdot 0.85 \cdot f_u \cdot (2 \cdot w \cdot t) \cdot \alpha \quad (7.5)$$

where $f_u$ is the tensile strength of the RHS, $w$ is defined in Equation (7.3), $t$ is the tube wall thickness, $\phi$ is the capacity factor taken as 0.9 and $\alpha$ is the shear lag reduction factor given as follows:

$$\alpha = 0.4 + 0.5 \cdot L_w / w \quad \text{when } 0.6 \cdot w \leq L_w \leq 1.0 \cdot w \quad (7.6a)$$

$$\alpha = 0.9 \quad \text{when } 1.0 \cdot w < L_w \leq 1.2 \cdot w \quad (7.6b)$$

$$\alpha = 1.0 \quad \text{when } L_w > 1.2 \cdot w \quad (7.6c)$$

It should be pointed out that the majority of the research on block shear tear-out failure was focused on bolted connections (Geschwindner 2004, Driver et al 2004). This failure mode is more common in the bolted connections because of the reduced area that results from the bolt holes. The design formulae for longitudinal fillet welds given in AS 4100 and BS 5950 Part 1 determine the strength of welds based on the weld throat failure. It can be proven (see Section 7.4.2.3) that for most cold-formed RHS the design capacity given in AS 4100 and BS 5950 Part 1 are always lower than that given by Equation (7.4). In other words the formulae for longitudinal fillet welds in AS 4100 and BS 5950 Part 1 give conservative prediction of tear-out design capacity for slotted gusset plate connections to RHS.

The shear lag reduction factors proposed by Korol (1996) for slotted gusset plate connections to RHS are plotted in Figure 7.6. It seems that the shear lag reduction factors given by Korol (1996) are conservative when applying to thin RHS connections. Tests are needed to further calibrate the shear lag reduction factors for thin RHS connections.

### *7.4.2 Design Rules*

#### *7.4.2.1 AS 4100*

Design rules for longitudinal fillet welds in RHS were derived (Zhao and Hancock 1994, 1995b, Zhao et al 1999) based on both parent metal strength as in AS/NZS 4600 and weld metal strength as in AS 4100. Only those based on weld metal strength are presented here since that is the approach adopted in AS 4100-1998.

The nominal capacity of the connection in tension ($N_t$) can be expressed as

$$N_t = 0.6 \cdot f_{uw} \cdot a \cdot L \quad (7.7)$$

where $f_{uw}$ is the weld metal strength, $a$ is the weld throat thickness and $L$ is the weld length. The term $0.6f_{uw}$ represents the design shear stress of the weld metal where the value of $f_{uw}$ is the nominal tensile strength of weld metal given in Table 7.3.

Table 7.3 Nominal tensile strength of weld metal (from Table 9.7.3.10(1) of AS 4100-1998)

| Manual metal arc electrode (AS/NZS 1553.1) | Submerged arc (AS 1858.1)<br>Flux cored arc (AS 2203)<br>Gas metal arc (AS/NZS 2717.1) | Nominal tensile strength of weld metal, $f_{uw}$ N/mm$^2$ |
|---|---|---|
| E41XX | W40X | 410 |
| E48XX | W50X | 480 |

AS 4100 does not specify the requirement of end return welds as shown in Figure 7.1 (b). However a lower capacity factor of $\phi$ of 0.7 is given for longitudinally welded RHS with a thickness less than 3 mm. This is based on the reliability analysis carried out by Zhao and Hancock (1995b) where tests without end return welds were also included. The design capacity of the connection is given by

$$\phi \cdot N_t = \phi \cdot 0.6 f_{uw} \cdot a \cdot L \tag{7.8a}$$

$$\phi = 0.8 \text{ when } t \geq 3 \text{ mm} \tag{7.8b}$$

$$\phi = 0.7 \text{ when } t < 3 \text{ mm} \tag{7.8c}$$

*7.4.2.2 BS 5950 Part 1*

Unlike AS 4100, the end return welds are specified as a requirement in BS 5950 Part 1: fillet welds finishing at the ends or sides of parts should be returned continuously around the corners for a distance at least twice the leg length ($s$) of the weld.

The longitudinal shear capacity $P_L$ per unit length of weld should be taken as:

$$P_L = p_w \cdot a \tag{7.9}$$

where $p_w$ is the design strength of a fillet weld and $a$ is the throat thickness of the weld.

The design strength $p_w$ of a fillet weld is given in Table 7.4 (from Table 37 of BS 5950 Part 1), corresponding to the electrode classification and the steel grade.

Table 7.4 Design strength of fillet welds $p_w$ (N/mm$^2$)

| Steel grade | Electrode classification (see Table 10 of BS 5950 Part 1) | | |
|---|---|---|---|
| | 35 | 42 | 50 |
| S275 | 220 | 220 | 220 |
| S355 | 220 | 250 | 250 |
| S460 | 220 | 250 | 280 |

For other types of electrode and/or other steel grades:

$p_w = 0.5 f_{uw}$ with $p_w \leq 0.55 f_u$ (7.10)

where $f_{uw}$ is the minimum tensile strength of the electrode as specified in the relevant product standard and $f_u$ is the specified minimum tensile strength of the parent metal.

*7.4.2.3 Comment on Weld Throat Failure*

In order to have the design capacity ($\phi N_t$) of longitudinal fillet welds given in Equation (7.8) less than the parent metal shear resistance $V_r$ given in Equation (7.4), the condition required is:

$$\frac{V_r}{\phi \cdot N_t} = \frac{0.9 \times 0.67 \times f_y \cdot t}{0.8 \times 0.6 \times f_{uw} \cdot a} > 1.0$$

$$\frac{f_y}{f_{uw}} > \frac{0.8 \times 0.6 \times a}{0.9 \times 0.67 \times t}$$

Assume the weld leg length ($s$) is the same as the tube thickness ($t$), i.e. the weld throat thickness ($a$) becomes $\frac{t}{\sqrt{2}}$. Therefore

$$\frac{f_y}{f_{uw}} > \frac{0.8 \times 0.6 \times \frac{t}{\sqrt{2}}}{0.9 \times 0.67 \times t} = 0.56$$

Similarly when comparing Equation (7.9) and Equation (7.4), the above condition becomes:

$$\frac{f_y}{f_{uw}} > 0.58$$

The nominal yield stress of commonly used cold-formed RHS varies from 228 N/mm$^2$ to 460 N/mm$^2$ as listed in Table 2.4. The nominal tensile strength of commonly used weld metal is 410 N/mm$^2$ and 480 N/mm$^2$ as listed in Table 7.3. The smallest ratio of $f_y$ to $f_{uw}$ is about 0.56 (= 228/410).Therefore for most cold-formed RHS the design capacity given in AS 4100 and BS 5950 Part 1 are always lower than that given by Equation (7.4). In other words the formulae for longitudinal fillet welds in AS 4100

and BS 5950 Part 1 can be used conservatively to design slotted gusset plate connections to RHS in tear-out failure.

### *7.4.3 Examples*

*7.4.3.1 Example 1*

A 10 mm plate – Grade 350 is fillet welded to a 75 × 75 × 2.5 RHS – Grade C350 as shown in Figure 7.7. The electrode used is W502. Assume the fillet weld is SP category and the weld size (leg length $s$) is 3 mm.

Check if the welded connection can carry a design tension force ($N^*$) of 50 kN if the weld length ($L_w$) is 40 mm as shown in Figure 7.7.

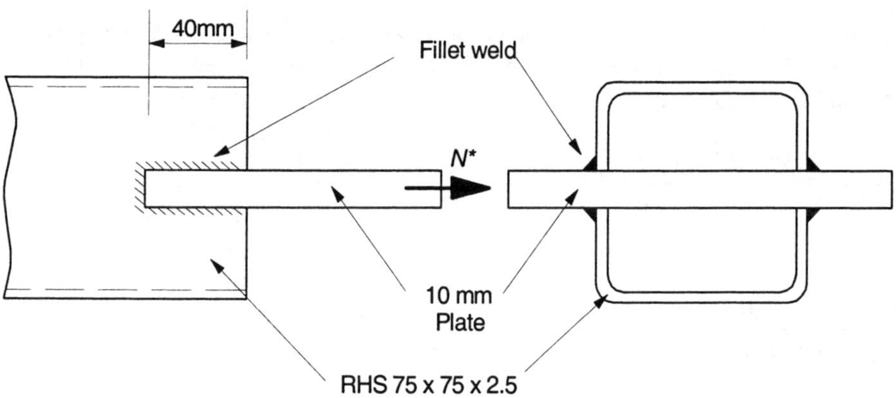


Figure 7.7 Slotted gusset-plate connection to cold-formed RHS (not to scale)

**Solution according to AS 4100**

1. Dimensions

$T = 10$ mm
$d = 75$ mm
$b = 75$ mm
$s = 3$ mm
$L_w = 40$ mm

2. Failure mode check

From Equation (7.3)

$w = d + b - T = 75 + 75 - 10 = 140$ mm

$L_w/w = 40/140 = 0.29 < 0.6$

The expected failure mode is tear-out failure. Equation (7.8) can be used for design as mentioned in Section 7.4.2.3.

3. Nominal capacity in tension

From Table 7.3, the nominal tensile strength of weld metal

$f_{uw} = 480$ N/mm$^2$

Weld throat thickness $a = \frac{s}{\sqrt{2}} = \frac{3}{\sqrt{2}} = 2.12$ mm

Total weld length (ignoring the end return welds) $L = 4 \times L_w = 4 \times 40 = 160$ mm

From Equation (7.7)

$$N_t = 0.6 \cdot f_{uw} \cdot a \cdot L = 0.6 \times 480 \times 2.12 \times 160 = 97{,}690 \text{ N} = 97.7 \text{ kN}$$

4. Design tension capacity check

From Equation (7.8c) $\phi = 0.7$

From Equation (7.8a)

$$\phi \cdot N_t = 0.7 \times 97.7 = 68.4 \text{ kN} > N^* = 50 \text{ kN}$$

Therefore the welded connection is satisfactory.

**Solution according to BS 5950 Part 1**

1. Dimensions

$T = 10$ mm
$d = 75$ mm
$b = 75$ mm
$s = 3$ mm
$L_w = 40$ mm

2. Failure mode check

From Equation (7.4)

$w = d + b - T = 75 + 75 - 10 = 140$ mm

$L_w/w = 40/140 = 0.29 < 0.6$

The expected failure mode is tear-out failure. Equation (7.9) can be used for design as mentioned in Section 7.4.2.3.

3. Design strength of fillet weld

Equation (7.10) should be used to determine $p_w$ because the steel grade C350 is not listed in Table 7.4.

$f_{uw} = 480$ N/mm$^2$ (from Table 7.3)
$f_u = 430$ N/mm$^2$ (from Table 2.4 in Chapter 2)

From Equation (7.10)

$p_w = 0.5f_{uw} = 0.5 \times 480 = 240$ N/mm$^2$
but $\leq 0.55f_u = 0.55 \times 430 = 237$ N/mm$^2$

Hence $p_w = 237$ N/mm$^2$

4. Longitudinal shear capacity per unit length of weld

Weld throat thickness $a = \frac{s}{\sqrt{2}} = \frac{3}{\sqrt{2}} = 2.12$ mm

From Equation (7.9)
$P_L = p_w \cdot a = 237 \times 2.12 = 502$ N/mm $= 0.5$ kN/mm

5. Design capacity check

Total weld length (ignoring the end return welds) $L = 4 \times L_w = 4 \times 40 = 160$ mm

The design stress ($f_w$), i.e. force per unit length transmitted by the fillet weld is

$f_w = N^*/L = 50$ kN/160 mm $= 0.313$ kN/mm $< P_L = 0.5$ kN/mm

Therefore the welded connection is satisfactory.

Note: It is interesting to note that the tension capacity (i.e. tension force carried by the total weld length) is checked in AS 4100 whereas the tension force per unit length is checked in BS 5950 Part 1. The equivalent tension capacity given by BS 5950 Part 1 would be $P_L L = 0.5 \times 160 = 80$ kN. This value of 80 kN is about 18% lower than the nominal capacity 97.7 kN predicted by AS 4100. However there is a capacity factor of 0.7 used in AS 4100 which leads to a design capacity of 68.4 kN. This makes the value given by BS 5950 (80 kN) about 17% higher than that of 68.4 kN given by AS 4100. It should be pointed out that slightly different load factors are used in AS 4100 and BS 5950 Part 1 as mentioned in Section 2.5.

#### *7.4.3.2 Example 2*

In Section 7.4.3.1 Example 1, what is the weld length ($L_w$) required in order to achieve the RHS tension capacity, i.e. without shear lag reduction?

**Solution**

In order to have the RHS fully yielded, the expected failure mode is RHS tension failure without shear lag reduction. According to Table 7.2 the required weld length ($L_w$) should be larger than 1.2$w$, i.e.

$L_w > 1.2\,w = 1.2 \times 140 = 168$ mm

*7.4.3.3 Example 3*

In Section 7.4.3.1 Example 1, what is the load carrying capacity if the weld length ($L_w$) is 100 mm?

**Solution**

1. Dimensions and Properties

$t$ =2.5 mm
$L_w$ = 100 mm
$w$ = 140 mm (from Section 7.4.3.1 Example 1)
$f_u$ = 430 N/mm$^2$ (from Table 2.4 of Chapter 2)

2. Failure mode check

$L_w/w = 100/140 = 0.714 > 0.6$

The expected failure mode is shear lag failure. Equation (7.5) can be used for design as mentioned in Section 7.4.1.

3. Shear lag reduction factor

When $0.6 \cdot w \leq L_w \leq 1.0 \cdot w$ the shear lag reduction factor is given by Equation (7.6a) as:

$$\alpha = 0.4 + 0.5 \cdot L_w / w = 0.4 + 0.5 \times 0.714 = 0.757$$

4. Shear lag resistance

From Equation (7.5)

$$T_r = \phi \cdot 0.85 \cdot f_u \cdot (2 \cdot w \cdot t) \cdot \alpha = 0.9 \times 0.85 \times 430 \times (2 \times 140 \times 2.5) \times 0.757 = 174{,}310 \text{ N}$$
$$\approx 174 \text{ kN}$$

Therefore the load carrying capacity is 174 kN if the weld length ($L_w$) is 100 mm.

## 7.5 Transverse Fillet Welds

### *7.5.1 Weld Failure in Shear*

To investigate welds in thin sections, tests were conducted by Zhao and Hancock (1995a) on cold-formed RHS ($t < 3$ mm) transversely welded to steel plates. The failure mode was found to be weld failure in shear as shown in Figure 7.8.

Figure 7.8 Weld failure in shear

### *7.5.2 Strength of Fillet Welds (Transverse Direction versus Longitudinal Direction)*

One of the parameters affecting the strength of fillet welds is the direction of loading. Two special cases are transverse fillet welds where the load is perpendicular to the welds, and longitudinal fillet welds where the load is parallel to the welds. In general the weld strength of transverse fillet welds is higher than that of longitudinal fillet welds mainly as a result of shear lag in longitudinal fillet welds. The comparisons of the two strengths are reported in Zhao and Hancock (1995a) based on existing theoretical and experimental studies in the literature. They are summarised in Tables 7.5 and 7.6 where $f_{wT}$ and $f_{wL}$ are forces carried by unit weld length for transverse and longitudinal fillet welds respectively. Details can be found in Zhao and Hancock (1993b).

It can be seen from Table 7.5 that the theoretical strength of the transverse fillet welds ranges from 17% to 50% higher than that of the longitudinal fillet welds, except for the case of vectorial addition where the strength of the transverse fillet welds and that of the longitudinal fillet welds are the same. The variation in the predictions of strength in different theories is most likely due to the different assumptions adopted.

It can be seen from Table 7.6 that the test strength of the transverse fillet welds ranges from 18% to 82% higher than that of the longitudinal fillet welds. The variation in values of strength in different tests is most likely due to differences in the parent metal, the weld metal and the types of testing.

Table 7.5 Comparison of theoretical values

| Theory type | $f_{wT}/f_{wL}$ | Reference |
|---|---|---|
| Vectorial addition | 1.00 | Mendelson (1968) |
| von Mises combination | 1.22 | Mendelson (1968) |
| Load-deformation response | 1.45 | Butler et al (1972) |
| Theory of elasticity | 1.47 | Kato and Morita (1974) |
| Ultimate load approach | 1.19 | Swannell (1981) |
| Theory of plasticity | 1.41 | Neis (1985) |
| Theory of equilibrium | 1.17 | Marsh (1988) |
| Lower-bound theory of plasticity | 1.41 | Kamtekar (1987) |
| Empirical equation | 1.50 | Lesik and Kennedy (1990) |

Table 7.6 Comparison of experimental values

| Investigators | $f_{wT}/f_{wL}$ |
|---|---|
| Higgins and Preece (1969) | 1.45 |
| Clarke (1970) and Clarke (1971) | 1.82 |
| Butler and Kulak (1971) | 1.42 |
| Kato and Morita (1974) | 1.56 |
| Swannell and Skewes (1979) | 1.18 |
| Marsh (1985) | 1.50 |
| Soetons (1987) | 1.19 |
| Marsh (1988) | 1.24 |
| Sanaei and Kamtekar (1988) | 1.18 |
| Miazga and Kennedy (1989) | 1.43 |

For fillet welds in thin cold-formed RHS members ($t \leq 3$ mm), similar comparisons are made in Figure 7.9. They are based on the test data in Zhao and Hancock (1993a, 1993b, 1994). It seems that in general the ratio reduces as the wall thickness of RHS increases. The ratio ranges from 1 to 2 with the lower bound ranging from 1 to 1.5 for fillet welds in thin cold-formed RHS members ($t \leq 3$ mm). The ratios recommended in BS 5950 Part 1 and AISC-LRFD (1999), i.e. 1.25 and 1.50 are also plotted in Figure 7.9. More discussions on the basis of increased fillet weld strength can be found in Iwankiw (1997).

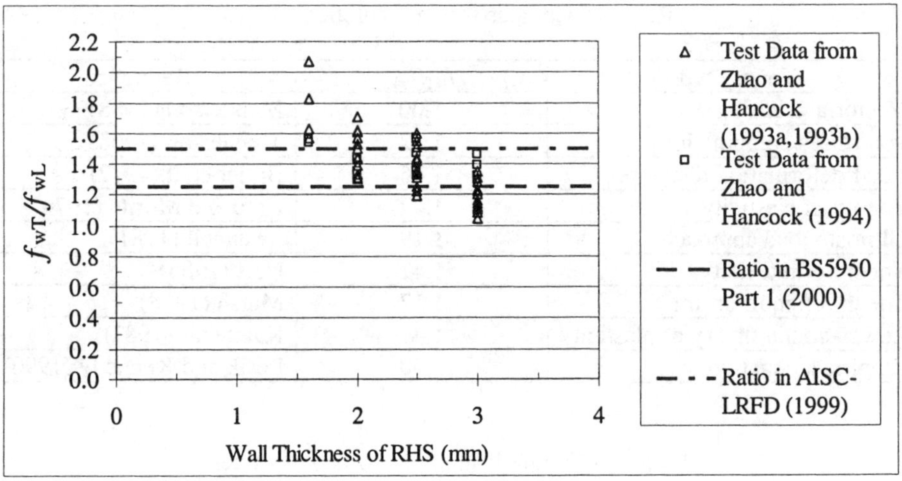


Figure 7.9 Strength ratio ($f_{wT}/f_{wL}$) versus wall thickness of RHS

### 7.5.3 *Design Rules*

*7.5.3.1 AS 4100*

Based on the test data shown in Figure 7.9 a proposed design rule was derived by Zhao and Hancock (1995a) for thin sections. The proposed nominal capacity of the connection ($t \leq 3$ mm) in tension ($N_t$) can be expressed as

$$N_t = 0.6 \cdot f_{uw} \cdot a \cdot L \cdot \left[0.75 + 0.5 \cdot \frac{t}{a}\right] \tag{7.11}$$

where $f_{uw}$ is the weld metal strength, $a$ is the weld throat thickness, $L$ is the weld length and $t$ is the thickness of the RHS.

Equation (7.11) takes into account the increase in strength in transverse fillet welds as discussed in Section 7.5.2. However this increase in strength has not been adopted in AS 4100, i.e. the same formulae for the longitudinal fillet welds are used for the transverse fillet welds except that only one capacity factor of 0.8 is used. They are repeated here as Equation (7.12).

$$N_t = 0.6 \cdot f_{uw} \cdot a \cdot L \tag{7.12a}$$

$$\phi \cdot N_t = \phi \cdot 0.6 f_{uw} \cdot a \cdot L \tag{7.12b}$$

$$\phi = 0.8 \text{ (from Table 3.4 of AS 4100)} \tag{7.12c}$$

*7.5.3.2 BS 5950 Part 1*

The transverse shear capacity $P_T$ per unit length of weld should be taken as:

$$P_T = 1.25 \cdot p_w \cdot a \tag{7.13}$$

where $p_w$ is the design strength of a fillet weld given in Table 7.4 or Equation (7.10) and $a$ is the weld throat thickness.

### *7.5.3.3 Comparison of Equivalent Weld Strength*

The expression of design capacity of transverse fillet welds (i.e. force carried by the total length of welds) is similar in AS 4100, BS 5950 Part 1 and AISC-LRFD. They all have a common term ($aL$) where $a$ is the throat thickness of the fillet weld and $L$ is the total weld length. Whatever left in the design capacity formulae apart from the term ($aL$) is call the equivalent weld strength in this chapter. The equivalent weld strengths in AS 4100, BS 5950 Part 1 and AISC-LRFD (1999) are compared in Table 7.7.

It can be seen from Table 7.7 that the equivalent weld strength given by BS 5950 Part 1 is 30% higher than that given by AS 4100, while the equivalent weld strength given by AISC-LRFD is about 40% higher than that given by AS 4100.

Table 7.7 Comparison of equivalent weld strength (transverse fillet weld only)

| Standard | AS 4100 | BS 5950 Part 1 (approximately) | AISC-LRFD |
|---|---|---|---|
| Formula | $\phi \cdot 0.6 \cdot f_{uw}$ | $1.25 \cdot 0.5 \cdot f_{uw}$ | $\phi \cdot 1.50 \cdot 0.6 \cdot f_{uw}$ |
| Capacity factor $\phi$ | 0.8 | N/A | 0.75 |
| Equivalent weld strength | $0.48 \cdot f_{uw}$ | $0.625 \cdot f_{uw}$ | $0.675 \cdot f_{uw}$ |
| % higher than that in AS 4100 | N/A | 30% | 41% |

### 7.5.4 *Examples*

A 75 × 75 × 2.5 RHS - Grade C350 is fillet welded to a 10 mm plate - Grade 350 using a W502 electrode. Because of access limitation, the fillet weld can only be applied to two sides of the section as shown in Figure 7.10. Assume the fillet weld is SP category and the weld size (leg length $s$) is 3 mm. Check if the welded connection can carry a design tension force ($N^*$) of 50 kN.

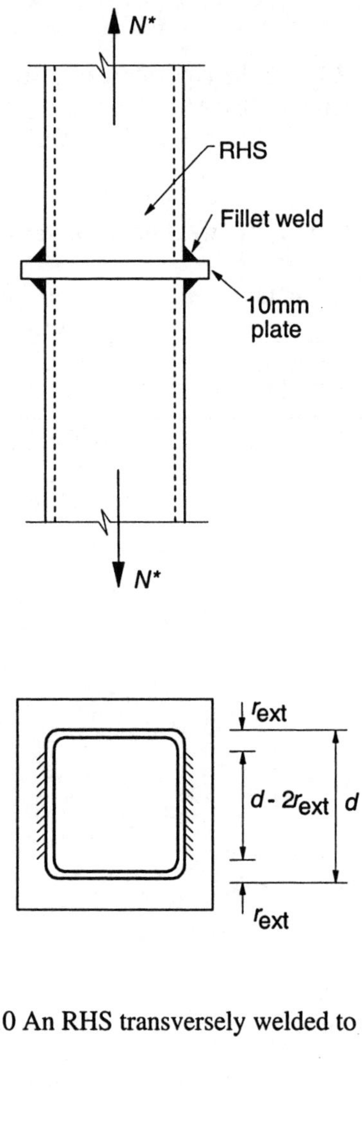


Figure 7.10 An RHS transversely welded to a steel plate

**Solutions according to AS 4100**

1. Dimensions

$d = 75$ mm
$b = 75$ mm
$t = 2.5$ mm

From Equation (6.1b), the external corner radius of the RHS
$r_{ext} = 2\,t = 2 \times 2.5 = 5$ mm

$s = 3$ mm

2. Nominal capacity in tension

From Table 7.3, the nominal tensile strength of weld metal

$f_{uw} = 480$ N/mm$^2$

Weld throat thickness $a = \dfrac{s}{\sqrt{2}} = \dfrac{3}{\sqrt{2}} = 2.12$ mm

Total weld length $L = 2 \cdot (d - 2 \cdot r_{ext}) = 2 \cdot (75 - 2 \times 5) = 130$ mm

From Equation (7.12a)

$N_t = 0.6 \cdot f_{uw} \cdot a \cdot L = 0.6 \times 480 \times 2.12 \times 130 = 79{,}373 \text{ N} = 79.4 \text{ kN}$

3. Design tension capacity check

From Equation (7.12c) $\phi = 0.8$

From Equation (7.12b)

$\phi \cdot N_t = 0.8 \times 79.4 = 63.5 \text{ kN} > N^* = 50 \text{ kN}$

Therefore the welded connection can carry a design tension force of 50 kN.

**Solutions according to BS 5950 Part 1**

1. Dimensions

$d = 75$ mm
$b = 75$ mm
$t = 2.5$ mm

From Equation (6.1b), the external corner radius of the RHS
$r_{ext} = 2\,t = 2 \times 2.5 = 5$ mm

$s = 3$ mm

2. Design strength of fillet weld

Equation (7.10) should be used to determine $p_w$ because the steel grade C350 is not listed in Table 7.4.

$f_{uw} = 480$ N/mm$^2$ (from Table 7.3)
$f_u = 430$ N/mm$^2$ (from Table 2.4 in Chapter 2)

From Equation (7.10)

$p_w = 0.5 f_{uw} = 0.5 \times 480 = 240$ N/mm$^2$
but $\leq 0.55 f_u = 0.55 \times 430 = 237$ N/mm$^2$

Hence $p_w = 237$ N/mm$^2$

3. Transverse shear capacity per unit length of weld

Weld throat thickness $a = \dfrac{s}{\sqrt{2}} = \dfrac{3}{\sqrt{2}} = 2.12$ mm

From Equation (7.13)
$P_T = 1.25 \cdot p_w \cdot a = 1.25 \times 237 \times 2.12 = 628$ N/mm $= 0.63$ kN/mm

4. Design capacity check

Total weld length $\quad L = 2 \cdot (d - 2 \cdot r_{ext}) = 2 \cdot (75 - 2 \times 5) = 130$ mm

The design stress ($f_w$), i.e. force per unit length transmitted by the fillet weld is

$f_w = N^*/L = 50$ kN/130 mm $= 0.39$ kN/mm $< P_T = 0.63$ kN/mm

Therefore the welded connection can carry a design tension force of 50 kN.

## 7.6 References


1. AISC (1999), Load and Resistance Factor Design Specification for Structural Steel Buildings, American Institute of Steel Construction, Chicago, Illinois, USA
2. AISI (1986), Specification for the Design of Cold-Formed Steel Structural Members, American Iron and Steel Institute, Washington D.C, USA
3. AWS (1998), Structural Welding Code – Sheet Steel, AWS D1.3-98, 4$^{th}$ Edition, American Welding Society, Miami, Florida, USA
4. AWS (2002), Structural Welding Code – Steel, AWS D1.1/D1.1M:2002, 18$^{th}$ Edition, American Welding Society, Miami, Florida, USA
5. BSI (2000), Structural use of Steelwork in Building, BS 5950, Part 1, British Standard Institution, London, UK
6. Butler, L.J. and Kulak, G.L. (1971), Strength of Fillet Welds as a Function of Direction of Load, *Welding Journal*, Welding Research Supplement, 50(5), pp 231-234
7. Butler, L.J., Pal, S. and Kulak, G.L. (1972), Eccentrically Loaded Welded Connections, *Journal of the Structural Division,* ASCE, 98(ST5), pp 989-1005
8. Cheng, J.J.R. & Kulak, G.L. (2000), Gusset Plate Connection to Round HSS Tension Members, *Engineering Journal,* American Institute of Steel Construction, 4$^{th}$ Quarter, pp 133-139
9. Clarke, A.H. (1970), The Strength of Fillet Welded Connections, *Master Thesis*, Civil Engineering Department, Imperial College, University of London, London, UK
10. Clarke, P.J. (1971), Basis of Design for Fillet Welded Joints under Static Loading, *Proceedings*, Conference on Improving Welding Product Design, The Welding Institute, Cambridge, Canada, pp 85-181
11. CSA-S16 (2001), Steel Structures for Buildings (Limit State Design), CSA-S16-01, Canadian Standards Association, Toronto, Ontario, Canada
12. CSA-S136 (2001), Cold-Formed Steel Structural Members, CSA-S136-01, Canadian Standards Association, Toronto, Ontario, Canada
13. Driver, R.G., Grondin, G.Y. and Kulak, G.L. (2004), A Unified Approach to Design for Block Shear, *Proceedings*, Fifth ECCS/AISC Workshop on Steel Connections, Session 6, Paper No. 4, Amsterdam, The Netherlands
14. EC3 (2003), Eurocode 3: Design of Steel Structures – Part 1.1: General Rules and Rules for Buildings, prEN 1993-1-1:2003, November 2003, European Committee for Standardization, Brussels, Belgium
15. EC3 (2004), Eurocode 3: Design of Steel Structures – Part 1.3: Supplementary Rules for Cold-Formed Members and Sheeting, EN 1993-1-3: 2004, March 2004, European Committee for Standardization, Brussels, Belgium
16. Funderburk, R. (2001), Selecting Filler Metals: Matching Strength Criteria, Welding Technology Institute of Australia, Sydney, Australia
17. Geschwindner, L.F. (2004), Evolution of Shear Lag and Block Shear Provisions in the AISC Specification, *Proceedings*, 5$^{th}$ ECCS/AISC Workshop on Steel Connections, Session 1, Paper No. 3, Amsterdam, The Netherlands
18. Higgins, T.R. and Preece, F.R. (1969), Proposed Working Stress for Fillet Welds in Building Construction, *Engineering Journal*, American Institute of Steel Construction, 6(1), pp 16-20
19. Iwankiw, N.R. (1997), Rational Basis for Increased Fillet Weld Strength, *Engineering Journal*, American Institute of Steel Construction, Second Quarter, pp 68-70

20. Kamtekar, A.G. (1987), The Strength of Inclined Fillet Welds, *Journal of Constructional Steel Research,* 7(1), pp 43-54
21. Kato, B. and Morita, K. (1974), Strength of Transverse Fillet Welded Joints, *Welding Journal,* Welding Research Supplement, 53(2), pp 49s-64s
22. Korol, R.M. (1996), Shear Lag in Slotted HSS Tension Members, *Canadian Journal of Civil Engineering,* Vol. 23, pp 1350-1354
23. Kyuba, H., Fukuda, Y. and Miki, C. (2001), Strength Matching of Butt-Welded Joint under Cyclic Loading, *Proceedings*, 1$^{st}$ International Conference on Steel and Composite Structures, Pusan, Korea, pp 261-266
24. Lesik, D.F. and Kennedy, D.J.L. (1990), Ultimate Strength of Fillet Welded Connections Loaded in Plane, *Canadian Journal of Civil Engineering,* Vol. 17, pp 55-67
25. Ligtenberg, F.K. (1968), International Test Series Final Report, *IIW Doc. XV-242-1968*, Stevin Laboratory, Delft University of Technology, Delft, The Netherlands
26. Ling, T.W., Zhao, X.L., Al-Mahaidi, R. and Packer, J.A. (2004), Connection Design of Very High Strength Steel Tubes Longitudinally Welded to Steel Plates, In: *Developments in Mechanics of Structures and Materials*, Deeks, A.J. and Hao, H. (eds), Balkema: London, pp 1135-1140
27. Marsh, C. (1985), Strength of Aluminum Fillet Welds, *Welding Journal,* Welding Research Supplement, 64(12), pp 335s-338s
28. Marsh, C. (1988), Strength of Aluminum T-joint Fillet Welds, *Welding Journal,* Welding Research Supplement, 67(8), pp 171s-176s
29. Mendelson, A. (1968), *Plasticity: Theory and Application*, The Macmillan Co., New York, USA
30. Miazga, G.S. and Kennedy, D.J.L. (1989), Behaviour of Fillet Welds as a Function of the Angle of Loading, *Canadian Journal of Civil Engineering,* Vol. 16, pp 583-599
31. NAS (2004), North American Specification for the Design of Cold-Formed Steel Structural Members, American Iron and Steel Institute, Washington D.C, USA
32. Neis, V.V. (1985), New Constitutive Law for Equal Leg Fillet Welds, *Journal of Structural Engineering*, ASCE, 111(8), pp 1747-1759
33. Packer, J.A. and Henderson, J.E. (1997), *Hollow Structural Section Connections and Trusses*, Canadian Institute of Steel Construction, Ontario, Canada
34. Peköz, T. and McGuire, W. (1981), Sheet Steel Welding, *Journal of the Structural Division,* ASCE, 107(ST8), pp 1657-1673
35. Pham, L. and Bennetts, I.D. (1983), Reliability of Fillet Weld Design, *Civil Engineering Transaction*, Institution of Engineers, Australia, E26(2), pp 119-124
36. Sanaei, E. and Kamtekar, A.G. (1988), Experimental on Some Arbitrarily Loaded Filet Welds, *Welding Journal,* Welding Research Supplement, 67(8), pp 103s-109s
37. Soetens, F. (1987), Welded Connections in Aluminum Alloy Structures, *Heron*, 32(1), pp 1-48
38. Standards Australia (1988), Cold-Formed Steel Structures, Australian Standard AS 1538, Standards Australia, Sydney, Australia
39. Standards Australia (1990), Steel Structures, Australian Standard AS 4100, Standards Australia, Sydney, Australia
40. Standards Australia (1990), Cored Electrodes for Arc-Welding - Ferritic Steel Electrodes, Australian Standard AS 2203.1, Standards Australia, Sydney, Australia

41. Standards Australia (1991), Structural Steel Welding – Part 1: Welding of steel structures, Australian Standard AS 1554.1, Standards Australia, Sydney, Australia
42. Standards Australia (1995), Covered Electrodes for Welding - Low Carbon Steel Electrodes for Manual Metal-Arc Welding of Carbon Steels and Carbon-Manganese Steels, Australian Standard AS/NZS 1553.1, Standards Australia, Sydney, Australia
43. Standards Australia (1996), Cold-Formed Steel Structures, Australian Standard AS/NZS 4600, Standards Australia, Sydney, Australia
44. Standards Australia (1996), Welding - Electrodes - Gas Metal Arc - Ferritic Steel Electrodes Australian Standard AS/NZS 2717.1, Standards Australia, Sydney, Australia
45. Standards Australia (1998), Steel Structures, Australian Standard AS 4100, Standards Australia, Sydney, Australia
46. Standards Australia (2000), Structural Steel Welding – Part 1: Welding of steel structures, Australian Standard AS/NZS 1554.1, Standards Australia, Sydney, Australia
47. Standards Australia (2003a), Methods for Destructive Testing of Welds in Metal - Macro metallographic test for cross-section examination, Australian Standard AS 2205.5.1, Standards Australia, Sydney, Australia
48. Standards Australia (2003b), Electrodes and Fluxes for Submerged-Arc Welding - Carbon Steels and Carbon-Manganese Steels, Australian Standard AS 1858.1, Standards Australia, Sydney, Australia
49. Stark, J.W.B. and Soetens, F. (1980), Welded Connections in Cold-Formed Sections, *Proceedings*, 5th International Specialty Conference on Cold-Formed Steel Structures, St. Louis, Missouri, USA, pp 591-636
50. Swannell, P. (1981), Rational Design of Fillet Weld Groups, *Journal of the Structural Division*, ASCE, 107(ST5), pp 789-802
51. Swannell, P. and Skewes, I.C. (1979), The Design of Welded Brackets Loaded in-plane: Elastic and Ultimate Load Techniques, *AWRA Report P6-8-77*, Australia Welding Research Association, Vol.7, pp 28-59
52. Wilkinson, T., Petrovski, T., Bechara, E. and Rubal, M. (2002), Experimental Investigation of Slot Lengths In RHS Bracing Members. *Advances in Steel Structures*, Chan, S.L., Teng, J.G. and Chung, K.F. (eds), Elsevier Science Ltd: Oxford, pp 205-212
53. Willibald, S., Packer, J.A., Martinez Saucedo, G. and Puthli, R.S. (2004), Shear Lag in Slotted Gusset Plate Connections to Tubes, *Proceedings*, Fifth ECCS/AISC Workshop on Steel Connections, Session 8, Paper No. 3, Amsterdam, The Netherlands
54. Zhao, X.L. and Hancock, G.J. (1993a), Tests and Design of Butt Welds and Transverse Fillet Welds in Thin Cold-Formed RHS Members, *Research Report, No. R681*, School of Civil and Mining Engineering, University of Sydney, Sydney, Australia
55. Zhao, X.L. and Hancock, G.J. (1993b), Tests and Design of Longitudinal Fillet Welds in Thin Cold-Formed RHS Members, *Research Report, No. R682*, School of Civil and Mining Engineering, University of Sydney, Sydney, Australia
56. Zhao, X.L. and Hancock, G.J. (1994), Tests and Design of Butt Welds and Fillet Welds in Cold-Formed DuraGal RHS Members, *Research Report, No. R702*, School of Civil and Mining Engineering, University of Sydney, Sydney, Australia

57. Zhao, X.L. and Hancock, G.J. (1995a), Butt Welds and Transverse Fillet Welds in Thin Cold-Formed RHS Members, *Journal of Structural Engineering*, ASCE, 121(11), pp 1674-1682.
58. Zhao, X.L. and Hancock, G.J. (1995b), Longitudinal Fillet Welds in Thin Cold-Formed RHS Member, *Journal of Structural Engineering*, ASCE, 121(11), pp 1683-1690
59. Zhao, X.L. and Hancock, G.J. (1996), Welded Connections in Thin Cold-Formed Rectangular Hollow Sections, In: *Connections in Steel Structures III*, Bjorhovde, R., Colson, A. and Zandonini, R. (eds), Pergamon: Oxford, pp 89-98
60. Zhao, X. L., Hancock, G.J. and Sully R.M. (1996), Design of Tubular Members and Connections Using Amendment No.3 to AS 4100, *Steel Construction*, Australian Institute of Steel Construction, 30(4), pp 2-15.
61. Zhao, X.L. and Hancock, G.J. (1998), Recent Research on Cold-Formed Tubular Structures, *Journal of Constructional Steel Research*, 46(1-3), paper 229, CD-Rom
62. Zhao, X.L., Al-Mahaidi, R. and Kiew, K.P. (1999), Longitudinal Fillet Welds in Thin-Walled C450 RHS Members, *Journal of Structural Engineering*, ASCE, 125(8), pp 821-828
63. Zhao, X.L. and Jiao, H. (2004), Recent Developments in High Strength Thin-Walled Steel Tubes, In: *Thin-Walled Structures – Recent Advances and Future Trends in Thin-Walled Structures Technology*, Loughlan, J. (ed), Canopus Publishing Limited: Bath, UK, pp 1-19

# Chapter 8: Welded Connections Subjected to Fatigue Loading

## 8.1 General

The fatigue phenomenon is characterised by a progressive degradation of strength under time variant stresses. The fatigue life (or fatigue resistance) of a structure includes a crack initiation phase and a crack propagation phase (Grundy 2004). The fatigue life of welded connections depends upon the connection detail and applied cyclic loading spectrum. Several methods can be used to determine the fatigue resistance of welded connections as summarised in Zhao et al (2000). They include the classification method, punching shear method, failure criterion method, static strength method, hot spot stress method and fracture mechanics method. Only the classification method is covered in this chapter since that is the method adopted in AS 4100 (Standards Australia 1998), Eurocode 3 Part 1.9 (EC 3 2003) and BS 7608 (BSI 1993). BS 5950 Part 1 (BSI 2000) recommends BS 7608 for fatigue design using the classification method. The hot spot stress method will be discussed in Section 9.2. The details of other methods may be found in Marshall (1992) for the punching shear method, in Mang and Bucak (1982) for the failure criterion method, in Kurobane (1989) and Niemi (1995) for the static strength method and Fisher et al (1970), Gurney (1979), Maddox (1991) and Zhao et al (1999) for the fracture mechanics method.

The fatigue design of hollow sections and simple connections is covered in Section 8.3 using AS 4100, Eurocode 3 Part 1.9 and BS 7608. The fatigue design of lattice girder joints is only covered in Eurocode 3 Part 1.9, which will be presented in Section 8.4.

Different symbols are used in various standards. The following symbols are adopted in this chapter.

D = Fatigue damage accumulation
$N_f$ = Number of stress cycles to failure
$N_i$ = Number of stress cycles to failure for a particular stress range $S_i$
$S_{DC}$ = Detail category
$S_{CA}$ = Constant Amplitude Fatigue Limit
$S_{CO}$ = Cut-Off Limit
$S_n$ = Nominal stress range
$S_i$ = One of the stress ranges (i) in a variable amplitude fatigue loading spectrum
m = Slope of $S_n$ - $N_f$ curve
$n_i$ = Number of stress cycles at a particular stress range $S_i$
$\phi$ = Capacity factor
$\gamma_{Mf}$ = Partial safety factor for fatigue strength

## 8.2 Classification Method

### *8.2.1 Design Procedures*

The classification method is based on structural details for different types of connections which are classified into various detail categories (also known as classes). Each detail category corresponds to a nominal stress range under which a connection will fail, with a given probability, after 2 million cycles. The design procedure can be summarised by the following steps:

Step 1 Determine the capacity factor or partial safety factor
Step 2 Determine the fatigue loading
Step 3 Determine the design stress ranges at the detail
Step 4 Determine the design spectrum on the detail
Step 5 Check for exemption from fatigue assessment
Step 6 Determine the appropriate detail category
Step 7 Determine the fatigue strength
Step 8 Check if fatigue limits are satisfied
Step 9 If not satisfactory modify the structure or application and repeat steps 1 to 8

Steps 2, 3 and 4 are related to fatigue loading. They are not covered in this chapter. The fatigue loading shall be obtained from the referring Standards such as AS1418 (Standards Australia 2002), BS5400 (BSI 1988) and Eurocode 1 (EC1 2002), where applicable. More descriptions can be found in the commentary of AS 4100 (Standards Australia 1999) and in BS 7608 (BSI 1993).

The other steps are described in this chapter.

### *8.2.2 Capacity Factor or Partial Safety Factor*

A capacity factor ($\phi$) is used in AS 4100 in determining the design fatigue strength. The capacity factor should not be taken as greater than 1.0. When the following conditions are satisfied (as given in Clause 11.1.6 in AS 4100), the capacity factor shall be taken as 1.0:

(a) The detail is located on a redundant load path, in a position where failure at that point alone will not lead to overall collapse of the structure.
(b) The stress history is estimated by conventional methods.
(c) The load cycles are not highly irregular.
(d) The detail is accessible for, and subject to, regular inspection.

The capacity factor shall be reduced when any of the above conditions do not apply. For non-redundant load paths, the capacity factor shall be less than or equal to 0.7. It is not practical to prescribe values of the capacity factor for the very wide range of possible circumstances, and it is the responsibility of the designer to determine suitable values for the particular circumstances under consideration (Standards Australia 1999).

Partial safety factors are used in BS 7608 and Eurocode 3 Part 1.9 for fatigue loading ($\gamma_{Ff}$) and for fatigue strength ($\gamma_{Mf}$). The value of $\gamma_{Ff}$ is taken as 1.0 while the value of

$\gamma_{Mf}$ depends on both the consequences of failure (low or high) and assessment method (e.g. damage tolerant method or safe life method in Eurocode 3 Part 1.9). The low consequence of failure is similar to the concept of redundant load path used in AS 4100, i.e. the failure of that component alone will not lead to overall collapse of the structure.

BS 7608 adopts 1.0 for $\gamma_{Mf}$ for cases of adequate structural redundancy. It is suggested that an additional factor on fatigue life be considered for cases of inadequate structural redundancy. In defining this factor on fatigue life account should be taken of the accessibility of the joint and the proposed degree of inspection as well as the consequence of failure. However no specific values are given in BS 7608.

The value of $\gamma_{Mf}$ varies from 1.0 to 1.35 in Eurocode 3 Part 1.9 as shown in Table 8.1. The damage tolerant method should provide an acceptable reliability that a structure will perform satisfactorily for its design life, provided that a prescribed inspection and maintenance regime for detecting and correcting fatigue damage (by repair cracks, strengthening of elements, etc) is implemented throughout the design life of the structure. The safe life method does not require regular in-service inspection for fatigue damage.

It should be noted that the partial safety factor ($\gamma_{Mf}$) is approximately equivalent to the reciprocal of the capacity factor ($\phi$). The largest $\gamma_{Mf}$ value of 1.35 corresponds to a value of $\phi$ of 0.74.

Table 8.1 Partial safety factor $\gamma_{Mf}$ for fatigue strength
(from Table 3.1 of Eurocode 3 Part 1.9 (2003))

| Assessment method | Consequence of failure | |
|---|---|---|
| | Low consequence | High consequence |
| Damage tolerant | 1.0 | 1.15 |
| Safe life | 1.15 | 1.35 |

### 8.2.3 *Exemption from Fatigue Assessment*

In AS 4100, fatigue assessment is not required for a member, connection or detail, if the design stress range ($S_n$) is less than $\phi \times 27$ N/mm$^2$ or if the number of stress cycles (n) is less than $2 \cdot 10^6 \cdot \left( \frac{\phi \cdot 36}{S_n} \right)^3$. This exemption identifies structures or structural elements in which the stress ranges and number of stress cycles are low enough that even the worst details given in the standard would be satisfied.

In AS 4100 and Eurocode 3 Part 1.9, stress ranges in the spectrum below the cut-off limit ($S_{CO}$ at 100 million cycles) may be ignored in the fatigue assessment. No such cut-off limit is specified in BS 7608.

### 8.2.4 Detail Categories (Classes)

The constructional details with descriptions and the corresponding detail categories for hollow sections and simple connections are given in AS 4100, Eurocode 3 Part 1.9 and BS 7608 whereas those for lattice girder joints are also given in Eurocode Part 1.9. They are reproduced as Tables 8.2, 8.3, 8.5 and 8.7 presented later in this chapter. It should be noted that the arrow in the construction detail indicates the direction of the applied stress range while the thick curved line perpendicular to the arrow indicates the fatigue crack. The detail category for hollow sections and simple connections varies from 36 $N/mm^2$ to 140 $N/mm^2$ in AS 4100 and Eurocode 3 Part 1.9, and from 43 $N/mm^2$ to 80 $N/mm^2$ in BS 7608. The detail category for lattice girder joints varies from 36 $N/mm^2$ to 90 $N/mm^2$ in Eurocode 3 Part 1.9. Different detail categories may be specified for the same constructional detail with different tube thickness. In AS 4100 and in Eurocode 3 Part 1.9, the detail category with tube thickness below 8 mm tends to be lower than that with tube thickness of 8 mm or above. The increase in category for butt welds reflects the narrower tolerances that can be achieved when assembling thicker tubes. This seems in contradiction with the commonly accepted rule for thickness correction, as for example given in BS 7608, see Section 8.3.2. However it should be pointed out that the thickness correction factor on stress range in BS 7608 only applies to hollow sections with a thickness greater than 16 mm whereas the details given in Table 8.2 and 8.3 are limited for hollow sections with a thickness less than 12.5 mm. In Eurocode 3 Part 1.9, the detail category for lattice girders depends heavily on the ratio of the chord thickness to the brace thickness ($t_0/t_i$). A greater detail category is specified for a larger $t_0/t_i$ ratio.

### 8.2.5 Fatigue Strength Curves ($S_n - N_f$ Curves)

The fatigue strength curves are usually called $S_n$ - $N_f$ curves where $S_n$ is the nominal stress range and $N_f$ is the corresponding number of cycles to failure. They are obtained by tests, usually under constant amplitude loading, followed by statistical analysis to define the design curve. They are often plotted on a log-log scale. A schematic $S_n$ - $N_f$ curve for hollow sections and simple connections is shown in Figure 8.1 where some of the major terms are defined, e.g. Detail Category ($S_{DC}$), Slope (m), Constant Amplitude Fatigue Limit ($S_{CA}$), Cut-Off Limit ($S_{CO}$). The part of the curve below $S_{CA}$ is a model to account for variable amplitude effects on fatigue damage accumulation. Similarly a schematic $S_n$ - $N_f$ curve for lattice girders is shown in Figure 8.2 with similar definitions.

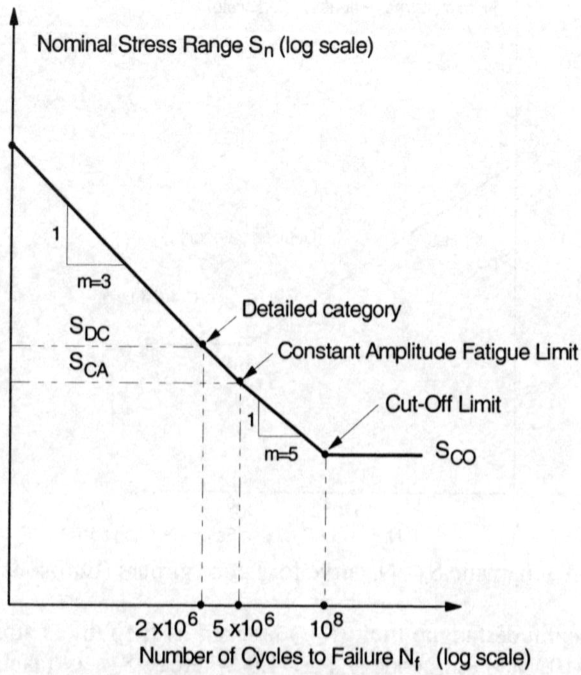


(a) AS 4100 and Eurocode 3 Part 1.9

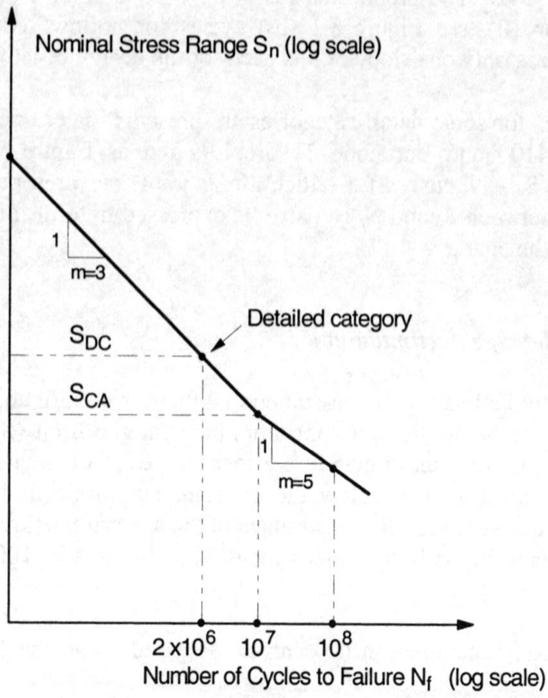


(b) BS 7608

Figure 8.1 A schematic $S_n$ - $N_f$ curve for hollow sections and simple connections

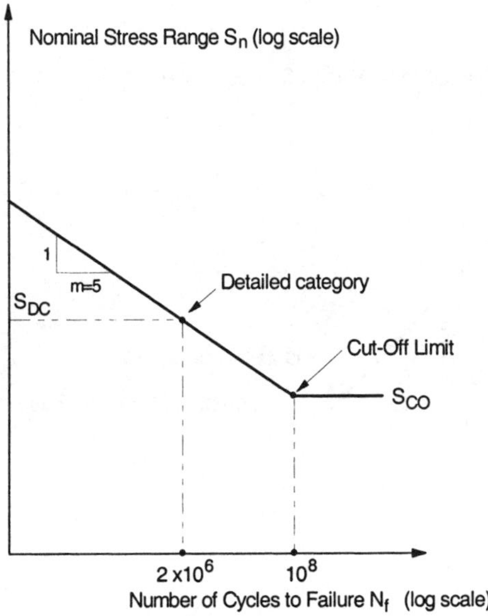


Figure 8.2 A schematic $S_n$ - $N_f$ curve for lattice girders (Eurocode 3 Part 1.9)

The constant amplitude fatigue limit ($S_{CA}$) is taken as the fatigue strength at $5 \times 10^6$ cycles in AS 4100 and Eurocode 3 Part 1.9 whereas $S_{CA}$ is taken as the fatigue strength at $10^7$ cycles in BS 7608. The cut-off limit ($S_{CO}$) is taken as the fatigue strength at $10^8$ cycles. The slope changes from 3 to 5 after $N_f$ of $5 \times 10^6$ (see Figure 8.1 (a)) or $10^7$ (see Figure 8.1 (b)) cycles for hollow sections and simple connections whereas only one slope of 5 is used for the design of lattice girders.

The $S_n$ - $N_f$ curves for some detail categories are presented later as Figure 8.4 of this chapter for AS 4100 and Eurocode 3 Part 1.9, and as Figure 8.5 for BS 7608 respectively. The $S_n$ - $N_f$ curves for lattice girder joints are presented in Figure 8.6. The relationship between $S_n$ and $N_f$ can also be expressed in terms of $S_{DC}$, $S_{CO}$ and m, as shown later in the chapter.

### 8.2.6 *Fatigue Damage Accumulation*

For constant amplitude loading, it is assumed that there is no fatigue damage when all the stress ranges are below the constant amplitude fatigue limit ($S_{CA}$). For variable amplitude loading, if any stress range in the spectrum exceeds the constant amplitude fatigue limit, the stress ranges below the constant amplitude limit should also be considered in the assessment. All stress ranges in the spectrum below the cut-off limit ($S_{CO}$) may be ignored in the fatigue assessment when using AS 4100 and Eurocode 3 Part 1.9.

The fatigue damage accumulation (D) can be assessed using the Palmgren-Miner's rule, i.e.

$$D = \sum \frac{n_i}{N_i} \tag{8.1}$$

in which $n_i$ is the number of cycles of a particular stress range $S_i$ and $N_i$ is the number of cycles to failure for that particular stress range. The allowable fatigue damage accumulation (D) for structures in a non-aggressive environment is generally taken as 1.0, if the effect of fatigue cracks and the possibility for inspection are taken into account by partial safety factors or capacity factors. This can be shown schematically in Figure 8.3.

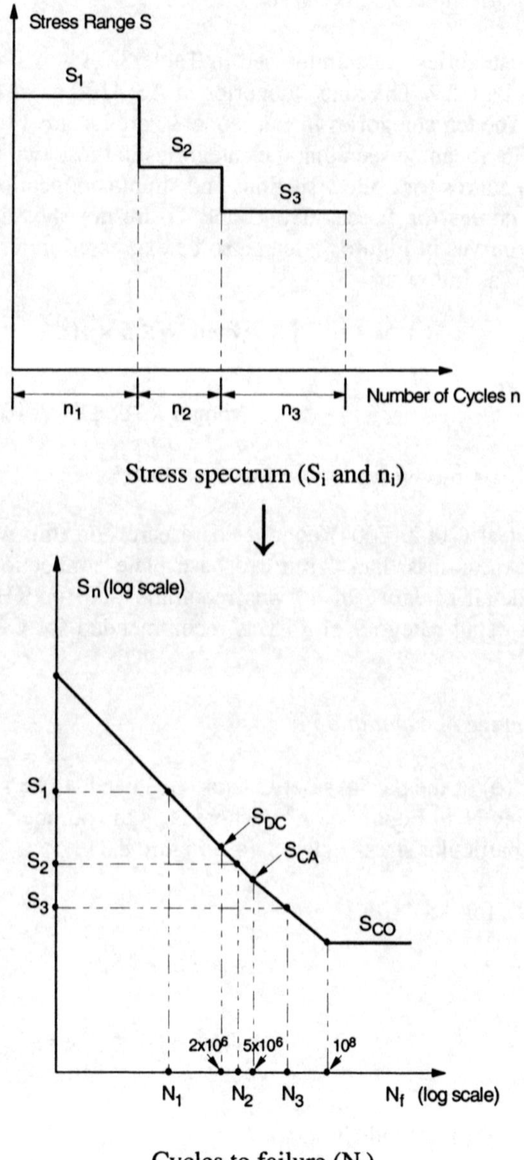


Check fatigue damage accumulation $D = \sum \frac{n_i}{N_i} = \frac{n_1}{N_1} + \frac{n_2}{N_2} + \frac{n_3}{N_3} < 1?$

Figure 8.3 Assessment of fatigue damage accumulation – a schematic view

## 8.3 Hollow Sections and Simple Connections

### *8.3.1 AS 4100 and Eurocode 3*

#### *8.3.1.1 Detail Categories and $S_n$ - $N_f$ Curves*

Most of the detail categories are summarised in Table 8.2 for AS 4100 and in Table 8.3 for Eurocode 3 Part 1.9. The nine categories in AS 4100 are 140, 90, 71, 56, 50, 45, 41, 40 and 36. The ten categories in Eurocode 3 Part 1.9 are 140, 125, 90, 71, 63, 56, 50, 45, 40 and 36. It can be seen that the categories in these two standards are very similar. The $S_n$ - $N_f$ curves for hollow sections and simple connections are reproduced in Figure 8.4. The curves for classes 63, 45 and 41 are not shown in Figure 8.4 for clarity. The $S_n$ - $N_f$ curves in Figure 8.4 can also be expressed in terms of $S_n$, $N_f$, $S_{DC}$, $S_{CA}$ and m (= 3 or 5) as follows:

$$S_n^3 = \frac{S_{DC}^3 \cdot 2 \cdot 10^6}{N_f} \qquad \text{when } N_f \leq 5 \times 10^6 \qquad (8.2a)$$

$$S_n^5 = \frac{S_{CA}^5 \cdot 5 \cdot 10^6}{N_f} \qquad \text{when } 5 \times 10^6 \leq N_f \leq 10^8 \qquad (8.2b)$$

in which, the values of $S_{DC}$ and $S_{CA}$ are listed in Table 8.4.

Recently Mashiri et al (2002, 2004) conducted research on thin-walled tubes (RHS and CHS with thickness less than 4 mm) to base plate connections under in-plane cyclic loading. A detail category of 40 was recommended for RHS to a base plate connection while a detail category of 45 was recommended for CHS to a base plate connection.

#### *8.3.1.2 Fatigue Damage Accumulation*

The capacity factor ($\phi$) or the partial safety factor ($\gamma_{Mf}$) needs to be applied to $S_{DC}$, $S_{CA}$ and $S_{CO}$ in calculating $N_i$ in Equation (8.1), where $N_i$ is the number of cycles to failure corresponding to a particular stress range $S_i$ (see Figure 8.3).

From Equation (8.2), for AS 4100

$$N_i = \frac{(\phi \cdot S_{DC})^3 \cdot 2 \cdot 10^6}{S_i^3} \qquad \text{when } S_i \geq \phi S_{CA} \qquad (8.3a)$$

$$N_i = \frac{(\phi \cdot S_{CA})^5 \cdot 5 \cdot 10^6}{S_i^5} \qquad \text{when } \phi S_{CO} \leq S_i < \phi S_{CA} \qquad (8.3b)$$

From Equation (8.2), for Eurocode 3 Part 1.9

$$N_i = \frac{(S_{DC} / \gamma_{Mf})^3 \cdot 2 \cdot 10^6}{S_i^3} \qquad \text{when } S_i \geq S_{CA}/\gamma_{Mf} \qquad (8.4a)$$

$$N_i = \frac{(S_{CA} / \gamma_{Mf})^5 \cdot 5 \cdot 10^6}{S_i^5} \qquad \text{when } S_{CO}/\gamma_{Mf} \leq S_i < S_{CA}/\gamma_{Mf} \qquad (8.4b)$$

Table 8.2 Detail categories for hollow sections and simple connections in AS 4100 (from Table 11.5.1(4) of AS 4100)

| Detail category | Constructional details | Description |
|---|---|---|
| 140 | 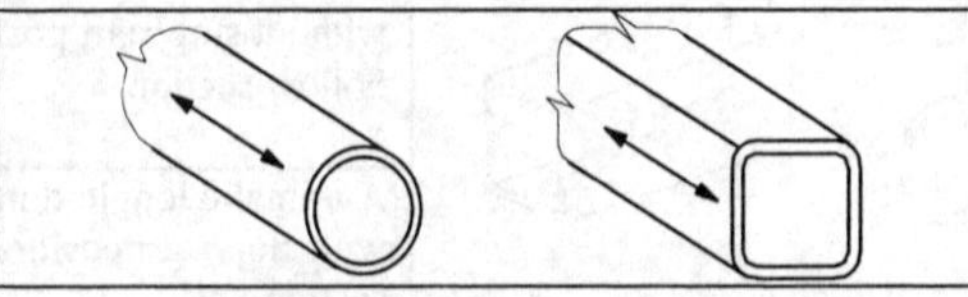 | Continuous automatic longitudinal welds: No stop-starts, or as manufactured. |
| 90 (t ≥ 8 mm)<br>71 (t < 8 mm) | 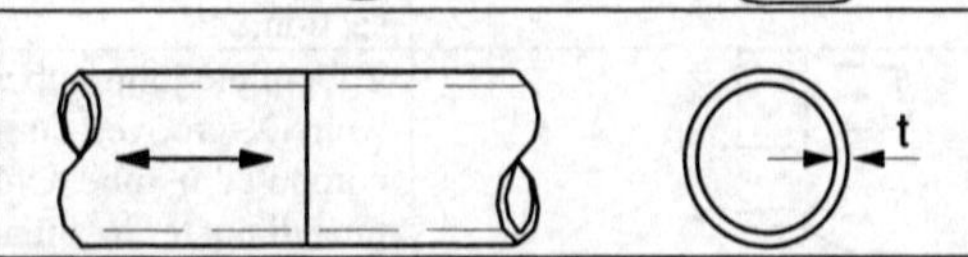 | Transverse butt welds: Butt-welded end-to-end connection of circular hollow sections. |
| 71 (t ≥ 8 mm)<br>56 (t < 8 mm) | 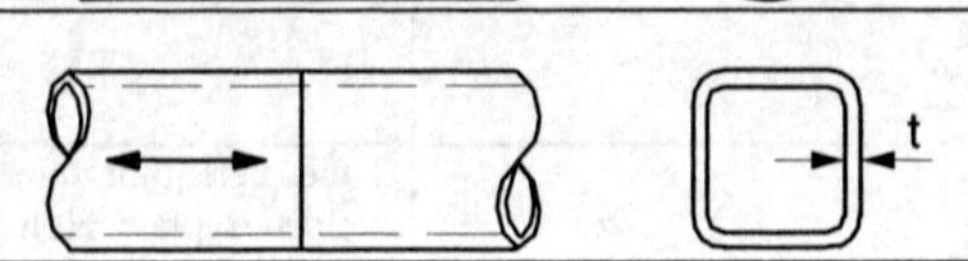 | Transverse butt welds: Butt-welded end-to-end connection of rectangular hollow sections. |
| 56 (t ≥ 8 mm)<br>50 (t < 8 mm) | 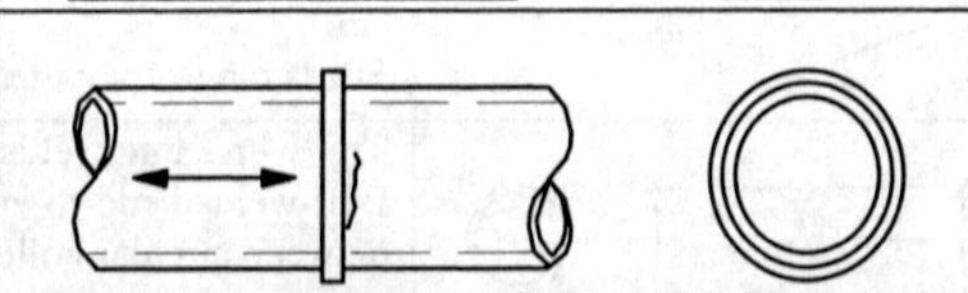 | Butt welds to intermediate plate: Circular hollow sections, end-to-end butt welded with an intermediate plate. |
| 50 (t ≥ 8 mm)<br>41 (t < 8 mm) | 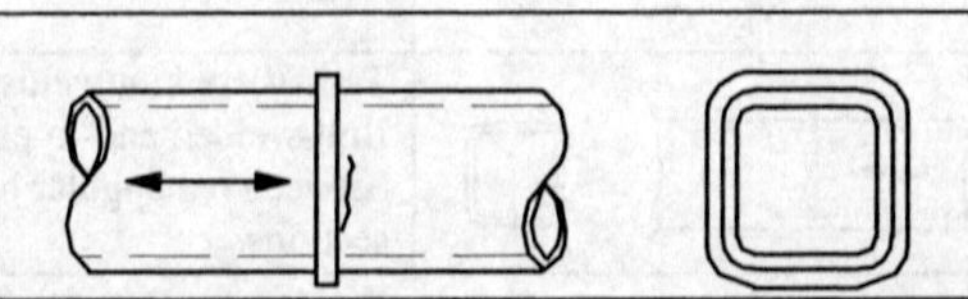 | Butt welds to intermediate plate: Rectangular hollow sections, end-to-end butt welded with an intermediate plate. |
| 71 | 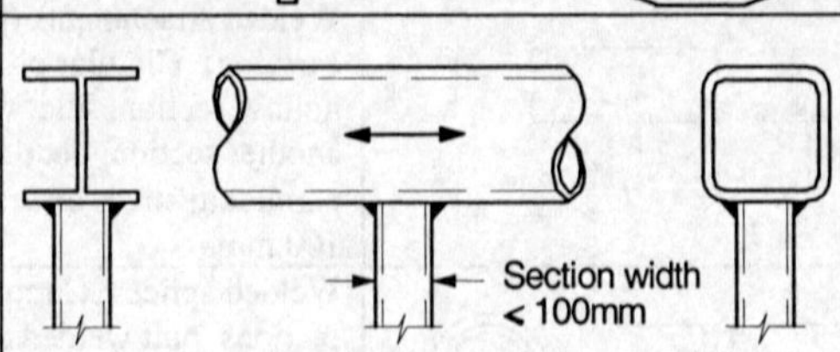 | Welded attachments (non-load-carrying): Circular or rectangular hollow section, fillet welded to another section. Section width parallel to stress direction ≤ 100 mm. |
| 45 (t ≥ 8 mm)<br>40 (t < 8 mm) | 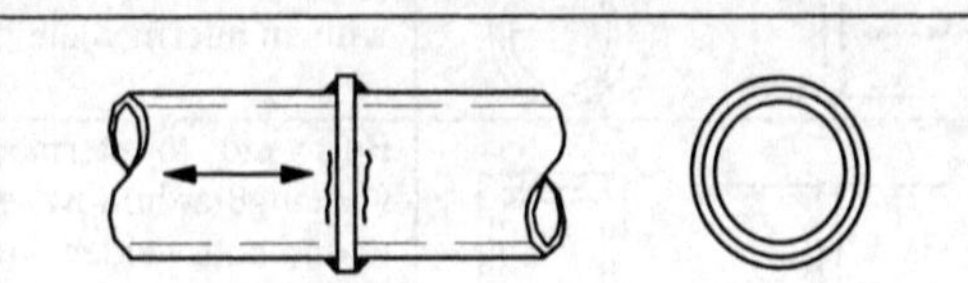 | Fillet welds to intermediate plate: Circular hollow sections, end-to-end fillet welded with an intermediate plate. |
| 40 (t ≥ 8 mm)<br>36 (t < 8 mm) | 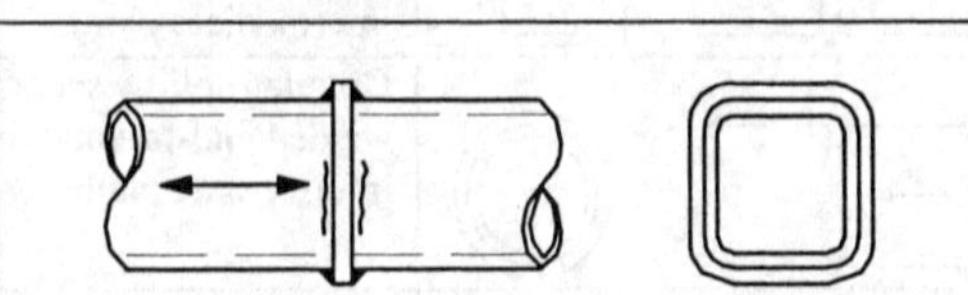 | Fillet welds to intermediate plate: Rectangular hollow sections, end-to-end fillet welded with an intermediate plate. |

Note: The arrow indicates the location and direction of the stresses acting in the basic material for which the stress range is to be calculated on a plane normal to the arrow.

Table 8.3 Detail categories for hollow sections and simple connections in Eurocode 3 Part 1.9 (from Table 8.2 and Table 8.6 of Eurocode 3 Part 1.9)

| Detail category | Constructional details | Description |
|---|---|---|
| 140 (t≤12.5 mm) | 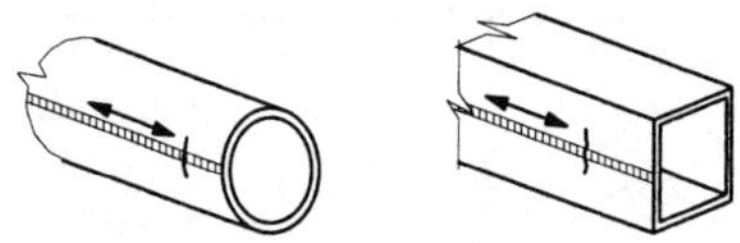 | Automatic longitudinal seam weld without stop/start positions in hollow section. |
| 125 (t>12.5 mm) | | |
| 90 (t>12.5 mm) | | Automatic longitudinal seam weld with stop/start positions in hollow section. |
| 71 (t ≤12.5 mm) | 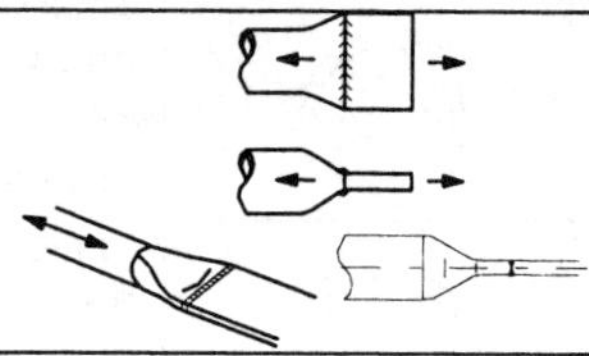 | Tube-plate joint, tubes flatted, butt weld (X-groove): stress range computed in tube. Only valid for tube diameter less than 200 mm. |
| 71 (α ≤45°) (t ≤12.5 mm) | 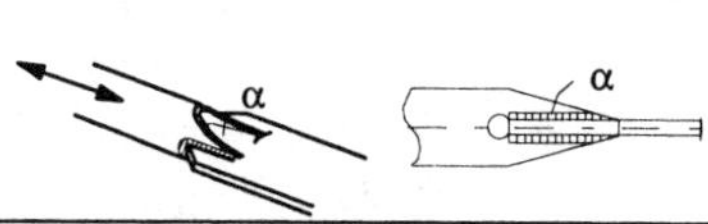 | Tube-plate joint, tube slit and welded to plate. Holes at end of slit. Stress range computed in tube. |
| 63 (α >45°) (t ≤12.5 mm) | | |
| 90 (8< t ≤12.5 mm) | 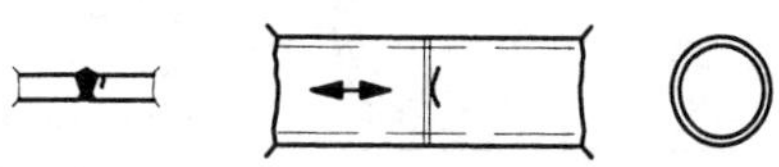 | Transverse butt welds: Butt-welded end-to-end connection between circular hollow sections. |
| 71 (t ≤ 8 mm) | | |
| 71 (8< t ≤12.5 mm) | 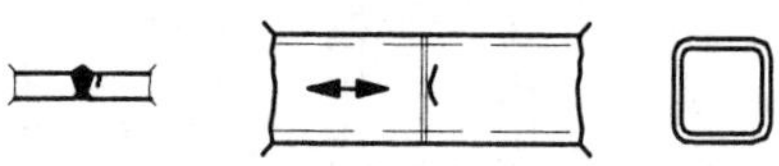 | Transverse butt welds: Butt-welded end-to-end connection between rectangular hollow sections. |
| 56 (t ≤ 8 mm) | | |
| 71 (t ≤12.5mm) | 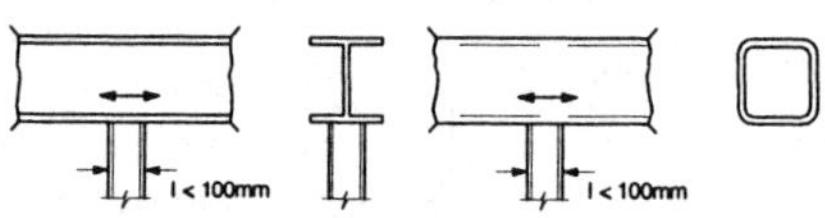  | Welded attachments (non-load-carrying): Circular or rectangular hollow section, fillet welded to another section. Section width parallel to stress direction ≤ 100 mm. |
| 56 (8< t ≤12.5 mm) | 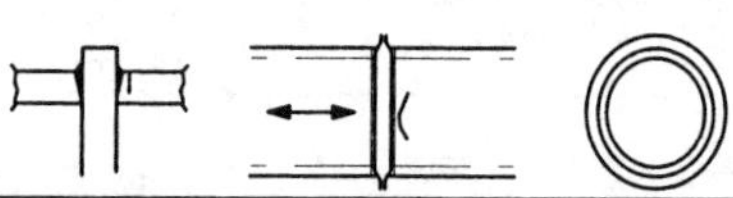 | Welded splices: Circular hollow sections, butt welded end-to-end with an intermediate plate. |
| 50 (t ≤ 8 mm) | | |
| 50 (8< t ≤12.5 mm) | 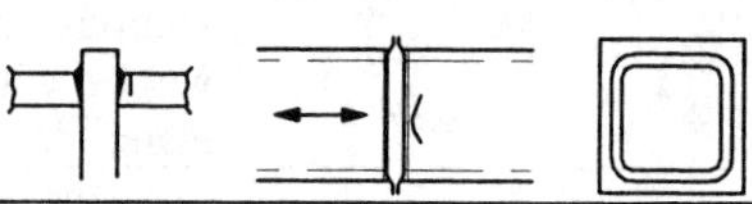 | Butt welds to intermediate plate: Rectangular hollow sections, end-to-end butt welded with an intermediate plate. |
| 45 (t ≤ 8 mm) | | |
| 40 (t ≤ 8 mm) | 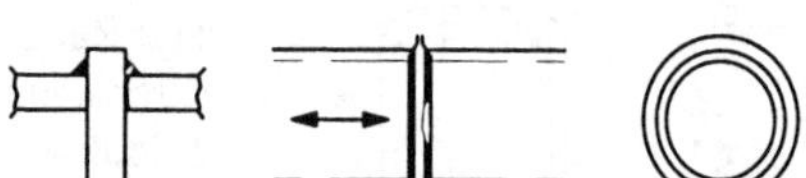 | Circular hollow sections, fillet welded end-to-end with an intermediate plate. |
| 36 (t ≤ 8 mm) |  | Rectangular hollow sections, fillet welded end-to-end with an intermediate plate. |

Note: The arrow indicates the location and direction of the stresses acting in the basic material for which the stress range is to be calculated on a plane normal to the arrow.

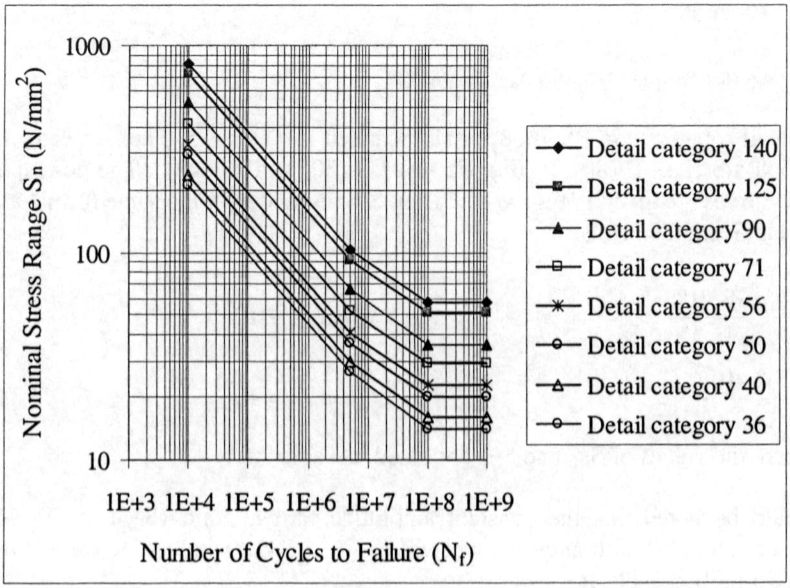


Figure 8.4 Some $S_n$ - $N_f$ curves for hollow sections and simple connections (AS 4100 and Eurocode 3 Part 1.9)

Table 8.4 Values of $S_{DC}$, $S_{CA}$ and $S_{CO}$ in AS 4100 and Eurocode 3 Part 1.9

| Detail Category $S_{DC}$ (N/mm$^2$) | Constant Amplitude Fatigue Limit $S_{CA}$ (N/mm$^2$) | Cut-off Limit $S_{CO}$ (N/mm$^2$) |
|---|---|---|
| 140 | 104 | 57 |
| 125 | 93 | 51 |
| 90 | 66 | 36 |
| 71 | 52 | 29 |
| 63 | 46 | 26 |
| 56 | 41 | 23 |
| 50 | 37 | 20 |
| 45 | 33 | 18 |
| 41 | 30 | 17 |
| 40 | 29 | 16 |
| 36 | 26 | 14 |

### 8.3.2 BS 7608

#### *8.3.2.1 Detail Categories and $S_n$ - $N_f$ Curves*

The detail categories in BS 7608 are summarised in Table 8.5. The five categories in BS 7608 are class E (80), F (68), F2 (60), G (50) and W (43). The corresponding $S_n$ - $N_f$ curves are given in Figure 8.5. They can be written in the same format as those shown in Section 8.3.1 as:

$$S_n^3 = \frac{S_{DC}^3 \cdot 2 \cdot 10^6}{N_f} \qquad \text{when } N_f \leq 10^7 \qquad (8.5a)$$

$$S_n^5 = \frac{S_{CA}^5 \cdot 10^7}{N_f} \qquad \text{when } N_f > 10^7 \qquad (8.5b)$$

in which, the values of $S_{DC}$ and $S_{CA}$ are listed in Table 8.6.

It should be noted that the constant amplitude fatigue limit ($S_{CA}$) in BS 7608 is defined at $N_f$ of 10 million cycles whereas $S_{CA}$ in AS4100 and Eurocode 3 is defined at $N_f$ of 5 million cycles.

#### *8.3.2.2 Fatigue Damage Accumulation*

Unlike Eurocode 3 Part 1.9, no partial safety factors ($\gamma_{Mf}$) are specified in BS 7608. It is suggested that the user may wish to set the design curve more or less than 2 standard deviations from the mean. For the purpose of comparison a partial safety factor ($\gamma_{Mf}$) is applied in this chapter to $S_{DC}$ and $S_{CA}$ in calculating $N_i$ in Equation (8.1), i.e.

$$N_i = \frac{(S_{DC} / \gamma_{Mf})^3 \cdot 2 \cdot 10^6}{S_i^3} \qquad \text{when } S_i \geq S_{CA}/\gamma_{Mf} \qquad (8.6a)$$

$$N_i = \frac{(S_{CA} / \gamma_{Mf})^5 \cdot 10^7}{S_i^5} \qquad \text{when } S_i < S_{CA}/\gamma_{Mf} \qquad (8.6b)$$

Table 8.5 Detail categories for hollow sections and simple connections in BS 7608 (from Table 7 and Table 10 of BS 7608)

| Class (Detail category) | Constructional details | Description |
|---|---|---|
| E (80) | 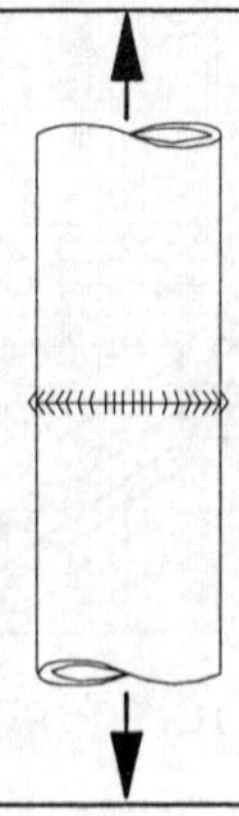 | Butt welded circular hollow sections: weld made from both sides. |
| F (68) | | Butt welded circular hollow sections: weld made from one side on permanent backing strip. |
| F2 (60) | | Butt welded circular hollow sections: weld made from one side with no backing strip. |
| F (68) | 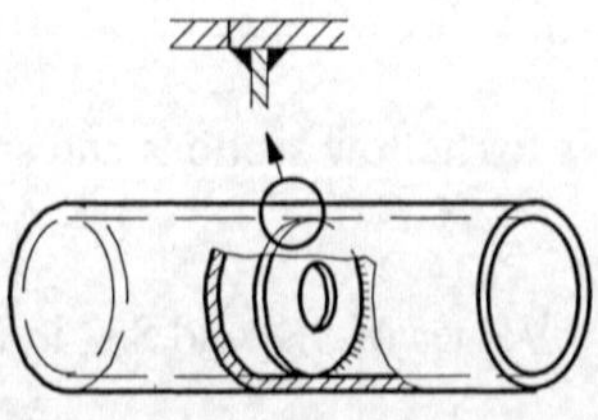 | Welded attachments: diaphragm or stiffener welded to circular hollow sections. Stress should include the stress concentration factor due to overall shape of adjoining structure. |
| F (68) (If toe-to-toe length ≤ 150 mm) | 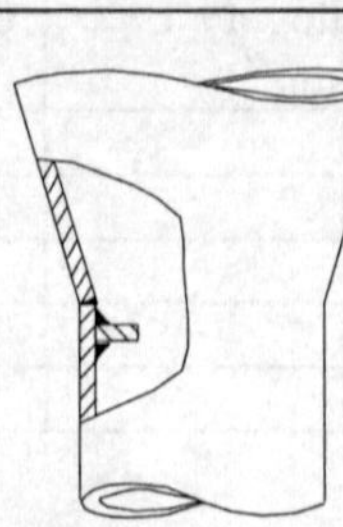 | Welded attachment: Potential crack in the stressed member at the toes of bevel butt or fillet welded attachment in a region of stress concentration. |
| F2 (60) (If toe-to-toe length > 150 mm | | |
| G (50) (if the attachment is a cover plate) | | |
| F (68) | 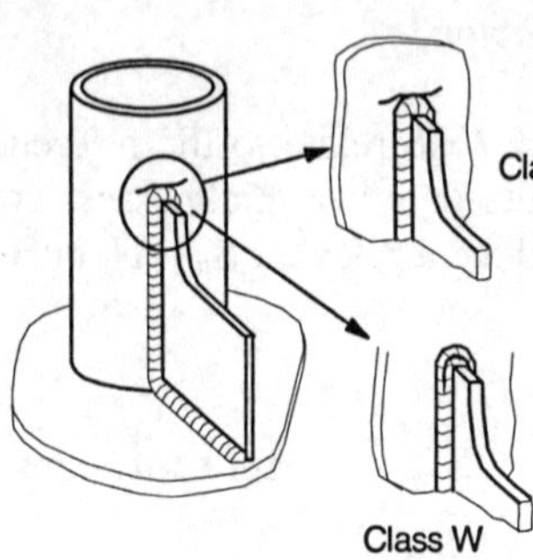 | Welded gusset connection: Potential crack in the tube at the toe of full penetration or fillet welded gusseted connections. The design stress has to include any local bending stress adjacent to the weld end. |
| W (43) | | Welded gusset connection: Weld throat failure in fillet welded gusseted connections. |

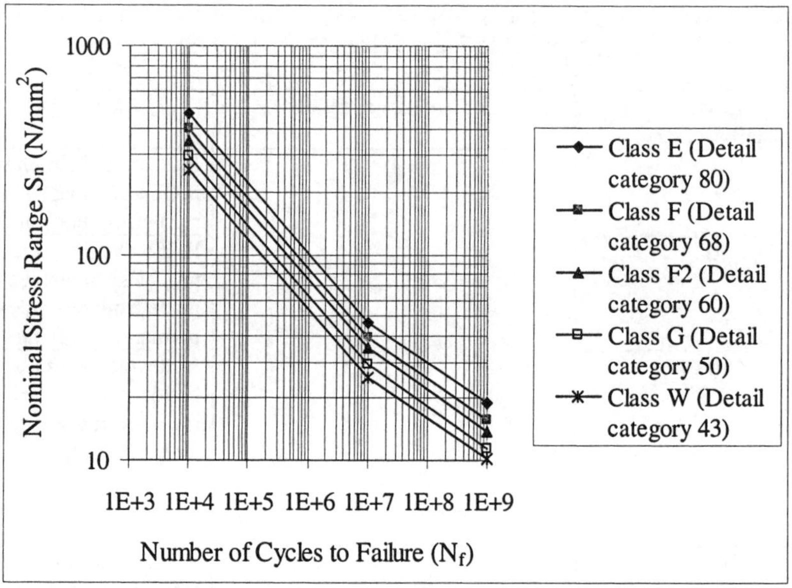


Figure 8.5 Some $S_n$ - $N_f$ curves for hollow sections and simple connections (BS 7608)

Table 8.6 Values of $S_{DC}$ and $S_{CA}$ in BS 7608

| Class in BS 7608 | Detail Category $S_{DC}$ (N/mm$^2$) | Constant Amplitude Fatigue Limit $S_{CA}$ (N/mm$^2$) |
|---|---|---|
| E | 80 | 47 |
| F | 68 | 40 |
| $F_2$ | 60 | 35 |
| G | 50 | 29 |
| W | 43 | 25 |

*8.3.2.3 Thickness Correction Factor*

The basic $S_n - N_f$ curves in BS 7608 relate to the reference thickness of 16 mm for welded joints. For welded joints of other thicknesses, correction factors on life or stress should be applied to produce a relevant $S_n - N_f$ curve. The correction on stress range is expressed as:

$$S = S_B \cdot \left(\frac{t_B}{t}\right)^{0.25} = S_B \cdot \left(\frac{16}{t}\right)^{0.25} \quad \text{for } t > 16 \text{ mm} \tag{8.7}$$

where S is the fatigue strength of the joint under consideration, $S_B$ is the fatigue strength of the joint using the basic $S_n - N_f$ curve, t is the greater of 16 mm or the actual thickness of the member under consideration and $t_B$ is the maximum thickness relevant to the basic $S_n - N_f$ curve, which is 16 mm for welded joints. It should be noted that the majority of cold-formed tubes have a thickness less than 16 mm where no correction is needed.

## 8.4 Lattice Girder Joints

### *8.4.1 Detail Categories and $S_n - N_f$ Curves*

Eurocode 3 Part 1.9 gives the detail categories for lattice girder joints. They are summarised in Table 8.7 where six categories are included, i.e. 90, 71, 56, 50, 45 and 36. The $S_n$ - $N_f$ curves for lattice girder joints are given in Figure 8.6. They can also be expressed in terms of $S_n$, $N_f$, $S_{DC}$ and m (= 5).

$$S_n^5 = \frac{S_{DC}^5 \cdot 2 \cdot 10^6}{N_f} \qquad \text{when } N_f \leq 10^8 \qquad (8.8)$$

in which, the values of $S_{DC}$ are listed in Table 8.8.

It should be pointed out that the validity ranges for the design of lattice girders in Eurocode 3 Part 1.9 are quite limited. A wider range of lattice girders is covered in CIDECT Design Guide No. 8 using the hot spot stress method (Zhao et al 2000).

### *8.4.2 Magnification Factors*

For lattice girders made of steel hollow sections the modeling for nominal stress ranges may be based on a simplified truss model with pinned connections. A magnification factor (MF) should be applied to the axial stress ranges to account for secondary bending moments. The value of the magnification factor is given in Table 8.9.

### *8.4.3 Fatigue Damage Accumulation*

The partial safety factor ($\gamma_{Mf}$) needs to be applied to $S_{DC}$ and $S_{CO}$ in calculating $N_i$ in Equation (8.1), where $N_i$ is the number of cycles to failure corresponding to a particular stress range $S_i$.

From Equation (8.8)

$$N_i = \frac{(S_{DC}/\gamma_{Mf})^5 \cdot 2 \cdot 10^6}{S_i^5} \qquad \text{when } S_i \geq S_{CO}/\gamma_{Mf} \qquad (8.9)$$

Table 8.7 Detail categories for lattice girder joints in Eurocode 3 Part 1.9 (from Table 8.7 of Eurocode 3 Part 1.9)

| Detail category | Constructional detail | Requirements |
|---|---|---|
| 90 $(\frac{t_0}{t_i} \geq 2.0)$ | Detail 1: CHS gap K and N joints 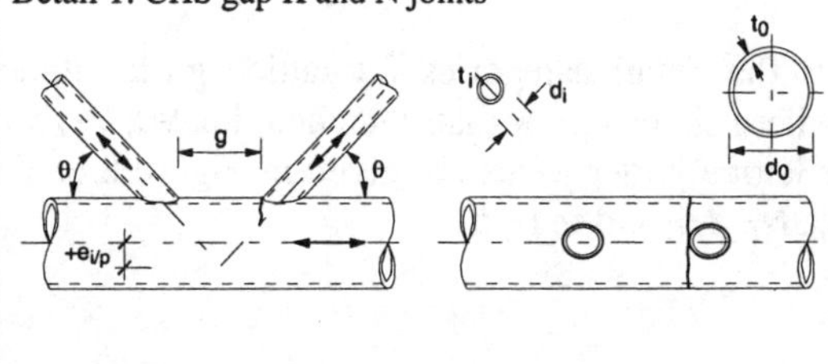 | Details 1 and 2: $t_0$ and $t_i \leq 8$ mm; $35^o \leq \theta \leq 50^o$; $b_0/t_0 \times t_0/t_i \leq 25$; $d_0/t_0 \times t_0/t_i \leq 25$; $0.4 \leq b_i/b_0 \leq 1.0$; $0.25 \leq d_i/d_0 \leq 1.0$; $b_0 \leq 200$ mm; $d_0 \leq 300$ mm; $-0.5h_0 \leq e_{i/p} \leq 0.25h_0$; $-0.5d_0 \leq e_{i/p} \leq 0.25d_0$; $e_{o/p} \leq 0.02b_0$ or $\leq 0.02d_0$ ($e_{o/p}$ is out-of-plane eccentricity) |
| 45 $(\frac{t_0}{t_i} = 1.0)$ | | |
| 71 $(\frac{t_0}{t_i} \geq 2.0)$ | Detail 2: RHS gap K and N joints 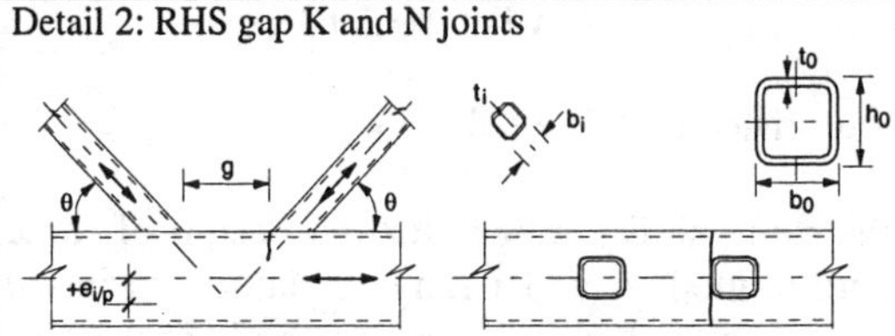 | Detail 2: $0.5(b_0-b_i) \leq g \leq 1.1(b_0-b_i)$ and $g \geq 2t_0$ |
| 36 $(\frac{t_0}{t_i} = 1.0)$ | | |
| 71 $(\frac{t_0}{t_i} \geq 1.4)$ | Detail 3: CHS or RHS overlap K joints 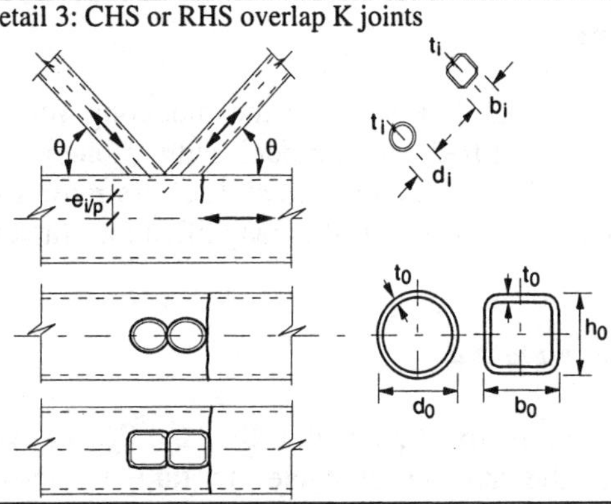 | Details 3 and 4: $30\% \leq$ overlap $\leq 100\%$; overlap = $(q/p) \times 100\%$; $t_0$ and $t_i \leq 8$ mm; $35^o \leq \theta \leq 50^o$; $b_0/t_0 \times t_0/t_i \leq 25$; $d_0/t_0 \times t_0/t_i \leq 25$; $0.4 \leq b_i/b_0 \leq 1.0$; $0.25 \leq d_i/d_0 \leq 1.0$; $b_0 \leq 200$ mm; $d_0 \leq 300$ mm; $-0.5h_0 \leq e_{i/p} \leq 0.25h_0$; $-0.5d_0 \leq e_{i/p} \leq 0.25d_0$; $e_{o/p} \leq 0.02b_0$ or $\leq 0.02d_0$ ($e_{o/p}$ is out-of-plane eccentricity) |
| 56 $(\frac{t_0}{t_i} = 1.0)$ | | |
| 71 $(\frac{t_0}{t_i} \geq 1.4)$ | Detail 4: CHS or RHS overlap N joints 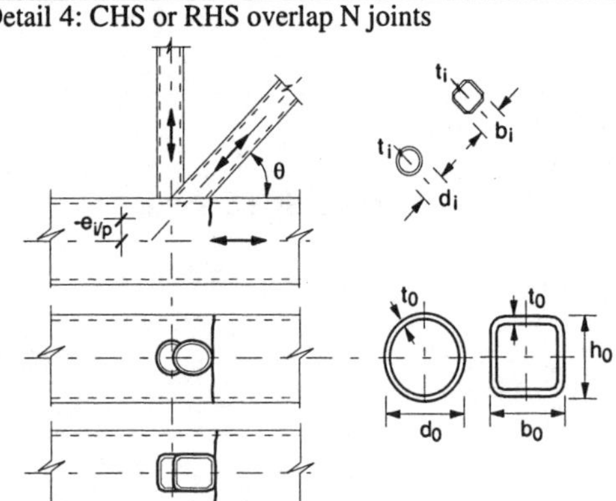 | Definition of p and q 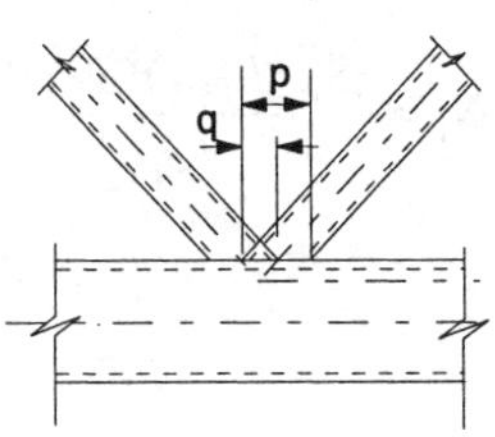 |
| 50 $(\frac{t_0}{t_i} = 1.0)$ | | |

General requirements:
Separate assessments needed for the chords and the braces;
For intermediate values of the ratio $t_0/t_i$ interpolate linearly between details categories;
Fillet welds permitted for braces with wall thickness $t \leq 8$ mm.

Note: The arrow indicates the location and direction of the stresses acting in the basic material for which the stress range is to be calculated on a plane normal to the arrow.

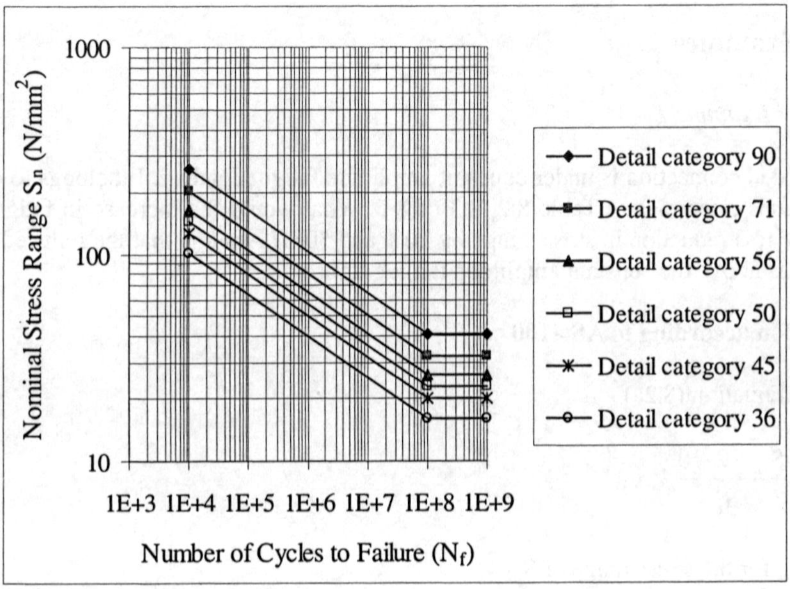


Figure 8.6 The $S_n$ - $N_f$ curves for lattice girder joints (Eurocode 3 Part 1.9)

Table 8.8 Values of $S_{DC}$ and $S_{CO}$ in Eurocode 3 Part 1.9 for lattice girder joints

| Detail Category $S_{DC}$ ($N/mm^2$) | Cut-off Limit $S_{CO}$ ($N/mm^2$) |
|---|---|
| 90 | 41 |
| 71 | 32 |
| 56 | 26 |
| 50 | 23 |
| 45 | 20 |
| 36 | 16 |

Table 8.9 Magnification factor (MF) to account for secondary bending moments in joints of lattice girders (from Table 4.1 and Table 4.2 of Eurocode 3 Part 1.9)

| Type of Joint | Chords | Vertical Braces | Diagonal Braces |
|---|---|---|---|
| CHS gap K-joint | 1.5 | 1.0 | 1.3 |
| CHS gap N-joint | 1.5 | 1.8 | 1.4 |
| CHS overlap K-joint | 1.5 | 1.0 | 1.2 |
| CHS overlap N-joint | 1.5 | 1.65 | 1.25 |
| RHS gap K-joint | 1.5 | 1.0 | 1.5 |
| RHS gap N-joint | 1.5 | 2.2 | 1.6 |
| RHS overlap K-joint | 1.5 | 1.0 | 1.3 |
| RHS overlap N-joint | 1.5 | 2.0 | 1.4 |

## 8.5 Examples

### *8.5.1 Example 1*

A welded connection is under constant amplitude fatigue loading. It belongs to one of the detail categories in Table 8.2, 8.3 or 8.5. What would the increase in fatigue life be due to a reduction in stress range by 20% and 50%? Assume that the reduced stress range is above the constant amplitude fatigue limit.

**Solution according to AS 4100**

From Equation (8.2a)

$$N_f = \frac{S_{DC}^3 \cdot 2 \cdot 10^6}{S_n^3}$$

Set $N_{f1}$ for the stress range of $S_{n1}$
Set $N_{f2}$ for the stress range of $S_{n2}$
Set $x$ is the reduction in stress range (e.g. $x = 0.20$ means 20% reduction)

If $S_{n2} = S_{n1}(1 - x)$

Then the ratio of $N_{f2}$ to $N_{f1}$ would become:

$$\frac{N_{f2}}{N_{f1}} = \frac{S_{n1}^3}{S_{n2}^3} = \frac{1}{(1-x)^3}$$

The relationship of $N_{f2}/N_{f1}$ and $x$ is shown in Figure 8.7. When $x = 0.20$, $N_{f2}/N_{f1} \approx 2$, and when $x = 0.50$, $N_{f2}/N_{f1} = 8$.

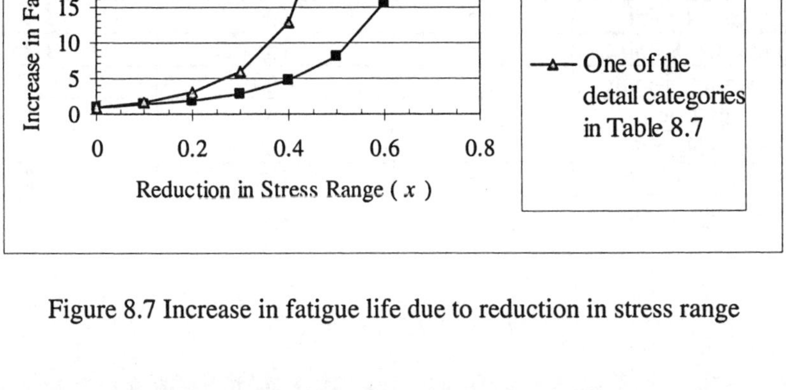


Figure 8.7 Increase in fatigue life due to reduction in stress range

**Solution according to BS 7608 and Eurocode 3 Part 1.9**

The solution is the same as that given by AS 4100 because BS 7608 and Eurocode 3 have the same slope m of 3 in the $S_n - N_f$ curves as in AS 4100.

### *8.5.2 Example 2*

A lattice girder joint is under constant amplitude fatigue loading. It belongs to one of the detail categories in Table 8.7. What would the increase in fatigue life be due to a reduction in stress range by 20% and 50%? Assume that the reduced stress range is above the cut-off limit.

**Solution according to Eurocode 3 Part 1.9**

From Equation (8.8)

$$N_f = \frac{S_{CA}^5 \cdot 2 \cdot 10^6}{S_n^5}$$

Set $N_{f1}$ for the stress range of $S_{n1}$
Set $N_{f2}$ for the stress range of $S_{n2}$
Set $x$ is the reduction in stress range (e.g. $x = 0.20$ means 20% reduction)

If $S_{n2} = S_{n1}(1 - x)$

Then the ratio of $N_{f2}$ to $N_{f1}$ would become:

$$\frac{N_{f2}}{N_{f1}} = \frac{S_{n1}^5}{S_{n2}^5} = \frac{1}{(1-x)^5}$$

The relationship of $N_{f2}/N_{f1}$ and $x$ is shown in Figure 8.7. When $x = 0.20$, $N_{f2}/N_{f1} \approx 3$, and when $x = 0.50$, $N_{f2}/N_{f1} = 32$.

It can be seen that the influence of stress range reduction on fatigue life is more beneficial for lattice girder joints because of the larger slope used in the $S_n - N_f$ curves for lattice girder joints.

### 8.5.3 *Example 3*

A butt-welded end-to-end connection between CHS is subjected to variable amplitude fatigue loading. The outer diameter of the CHS is 101.6 mm with a wall thickness of 6.4 mm. The weld is made from one side on permanent backing strip. The expected stress ranges ($S_i$) and corresponding estimated number of cycles ($n_i$) are shown below.

| Stress range number (i) | Expected stress ranges ($S_i$) N/mm$^2$ | Corresponding estimated number of cycles ($n_i$) |
|---|---|---|
| 1 | 80 | 200,000 |
| 2 | 60 | 100,000 |
| 3 | 35 | 1,000,000 |
| 4 | 20 | 400,000 |

Check the fatigue damage accumulation. Assume that the detail is located on a non-redundant load path and the consequence of failure is considered as high.

**Solution according to AS 4100**

1. Determine the capacity factor

From Section 8.2.2, the capacity factor $\phi = 0.7$ for non-redundant load path.

2. Determine the appropriate detail category

From Table 8.2, the detail category should be 71 for butt-welded end-to-end connection of CHS with $t < 8$ mm.

From Table 8.4, $S_{DC} = 71$ N/mm$^2$, $S_{CA} = 52$ N/mm$^2$ and $S_{CO} = 29$ N/mm$^2$

3. Check low stress range

The given stress range No.4 ($S_4 = 20$ N/mm$^2$) is less than the cut-off limit of $S_{CO}$ of 29 N/mm$^2$. Therefore this stress range will not be included in the calculations below.

4. Determine the fatigue strength $N_i$ at the stress range of $S_i$

$\phi S_{CA} = 0.7 \times 52 \approx 36$ N/mm$^2$
$\phi S_{CO} = 0.7 \times 29 \approx 20$ N/mm$^2$

Since $S_1$ and $S_2$ are larger than $\phi S_{CA}$, use Equation (8.3a)

$$N_1 = \frac{(\phi \cdot S_{DC})^3 \cdot 2 \cdot 10^6}{S_1^3} = \frac{(0.7 \cdot 71)^3 \cdot 2 \cdot 10^6}{80^3} = 0.48 \cdot 10^6$$

$$N_2 = \frac{(\phi \cdot S_{DC})^3 \cdot 2 \cdot 10^6}{S_2^3} = \frac{(0.7 \cdot 71)^3 \cdot 2 \cdot 10^6}{60^3} = 1.14 \cdot 10^6$$

Since $\phi S_{CO} \leq S_3 < \phi S_{CA}$, use Equation (8.3b)

$$N_3 = \frac{(\phi \cdot S_{CA})^5 \cdot 5 \cdot 10^6}{S_3^5} = \frac{(0.7 \times 52)^5 \cdot 5 \cdot 10^6}{35^5} = 6.08 \cdot 10^6$$

5. Check fatigue damage accumulation

$$D = \sum_{i=1}^{3} \frac{n_i}{N_i} = \frac{n_1}{N_1} + \frac{n_2}{N_2} + \frac{n_3}{N_3} = \frac{0.2 \cdot 10^6}{0.48 \cdot 10^6} + \frac{0.1 \cdot 10^6}{1.14 \cdot 10^6} + \frac{1.0 \cdot 10^6}{6.08 \cdot 10^6}$$
$$= 0.42 + 0.09 + 0.16 = 0.67$$

$D = 0.67 < 1.0$, satisfactory.

**Solution according to Eurocode 3 Part 1.9**

The solution according to Eurocode 3 Part 1.9 is the same as that given in AS 4100 except that a partial safety factor ($\gamma_{Mf}$) is used instead of a capacity factor.

From Table 8.1 with given conditions (non-redundant load path and high consequence of failure), $\gamma_{Mf} = 1.35$.

$S_{CA}/\gamma_{Mf} = 52/1.35 \approx 39$ N/mm$^2$
$S_{CO}/\gamma_{Mf} = 29/1.35 \approx 22$ N/mm$^2$

Since $S_1$ and $S_2$ are larger than $S_{CA}/\gamma_{Mf}$, use Equation (8.4a)

$$N_1 = \frac{(S_{DC}/\gamma_{Mf})^3 \cdot 2 \cdot 10^6}{S_1^3} = \frac{(71/1.35)^3 \cdot 2 \cdot 10^6}{80^3} = 0.57 \cdot 10^6$$

$$N_2 = \frac{(S_{DC}/\gamma_{Mf})^3 \cdot 2 \cdot 10^6}{S_2^3} = \frac{(71/1.35)^3 \cdot 2 \cdot 10^6}{60^3} = 1.35 \cdot 10^6$$

Since $S_{CO}/\gamma_{Mf} \leq S_3 < S_{CA}/\gamma_{Mf}$, use Equation (8.4b)

$$N_3 = \frac{(S_{CA}/\gamma_{Mf})^5 \cdot 5 \cdot 10^6}{S_3^5} = \frac{(52/1.35)^5 \cdot 5 \cdot 10^6}{35^5} = 8.07 \cdot 10^6$$

Check fatigue damage accumulation

$$D = \sum_{i=1}^{3} \frac{n_i}{N_i} = \frac{n_1}{N_1} + \frac{n_2}{N_2} + \frac{n_3}{N_3} = \frac{0.2 \cdot 10^6}{0.57 \cdot 10^6} + \frac{0.1 \cdot 10^6}{1.35 \cdot 10^6} + \frac{1.0 \cdot 10^6}{8.07 \cdot 10^6}$$
$$= 0.35 + 0.07 + 0.12 = 0.54$$

$D = 0.54 < 1.0$, satisfactory.

**Solution according to BS 7608**

1. Determine the partial safety factor

BS 7608 adopts 1.0 for $\gamma_{Mf}$ for cases of adequate structural redundancy. It is suggested that an additional factor on fatigue life be considered for cases of inadequate structural redundancy. In defining this factor on fatigue life account should be taken of the accessibility of the joint and the proposed degree of inspection as well as the consequence of failure. However no specific values are given in BS 7608. The value given in Eurocode 3 Part 1.9 is used in the calculation for the purpose of comparison.

From Table 8.1 with given conditions (non-redundant load path and high consequence of failure), $\gamma_{Mf} = 1.35$.

2. Determine the appropriate detail category

From Table 8.5, the detail category should be Class F for weld made from one side on permanent backing strip.

From Table 8.6, $S_{DC} = 68$ N/mm$^2$ and $S_{CA} = 40$ N/mm$^2$

3. Determine the fatigue strength $N_i$ at the stress range of $S_i$

$S_{CA}/\gamma_{Mf} = 40/1.35 \approx 30$ N/mm$^2$

Since $S_1$, $S_2$ and $S_3$ are all larger than $S_{CA}/\gamma_{Mf}$, use Equation (8.6a)

$$N_1 = \frac{(S_{DC}/\gamma_{Mf})^3 \cdot 2 \cdot 10^6}{S_1^3} = \frac{(68/1.35)^3 \cdot 2 \cdot 10^6}{80^3} = 0.50 \cdot 10^6$$

$$N_2 = \frac{(S_{DC}/\gamma_{Mf})^3 \cdot 2 \cdot 10^6}{S_2^3} = \frac{(68/1.35)^3 \cdot 2 \cdot 10^6}{60^3} = 1.18 \cdot 10^6$$

$$N_3 = \frac{(S_{DC}/\gamma_{Mf})^3 \cdot 2 \cdot 10^6}{S_3^3} = \frac{(68/1.35)^3 \cdot 2 \cdot 10^6}{35^3} = 5.96 \cdot 10^6$$

Since $S_4$ is less than $S_{CA}/\gamma_{Mf}$, use Equation (8.6b)

$$N_4 = \frac{(S_{CA}/\gamma_{Mf})^5 \cdot 10^7}{S_4^5} = \frac{(40/1.35)^5 \cdot 10^7}{20^5} = 71.4 \cdot 10^6$$

4. Check fatigue damage accumulation

$$D = \sum_{i=1}^{4} \frac{n_i}{N_i} = \frac{n_1}{N_1} + \frac{n_2}{N_2} + \frac{n_3}{N_3} + \frac{n_4}{N_4} = \frac{0.2 \cdot 10^6}{0.50 \cdot 10^6} + \frac{0.1 \cdot 10^6}{1.18 \cdot 10^6} + \frac{1.0 \cdot 10^6}{5.96 \cdot 10^6} + \frac{0.4 \cdot 10^6}{71.4 \cdot 10^6}$$
$$= 0.40 + 0.09 + 0.17 + 0.0056 = 0.67$$

$D = 0.67 < 1.0$, satisfactory.

Note: The fatigue damage accumulation (D) calculated using AS 4100, Eurocode 3 Part 1.9 and BS 7608 is very close ranging from 0.54 to 0.67. A smaller D would be obtained if a smaller partial safety factor $\gamma_{Mf}$ was adopted when using BS 7608. It can be proven that D would become 0.25 and 0.40 if the partial safety factor $\gamma_{Mf}$ was taken as 1.0 and 1.15 respectively.

### *8.5.4 Example 4*

For the same connection detail described in Section 8.5.3 Example 3, the daily stress cycles are listed below.

| Stress range number (i) | Expected stress ranges ($S_i$) N/mm$^2$ | Number of cycles per day ($n_{i,pd}$) |
|---|---|---|
| 1 | 80 | 30 |
| 2 | 60 | 100 |
| 3 | 35 | 1000 |
| 4 | 20 | 2000 |

Determine the fatigue life of the detail.

**Solution according to AS 4100**

1. Determine the fatigue strength $N_i$ at the stress range of $S_i$

From the solution of Section 8.5.3 Example 3

$$N_1 = 0.48 \cdot 10^6$$
$$N_2 = 1.14 \cdot 10^6$$
$$N_3 = 6.08 \cdot 10^6$$

2. Determine the number of cycles ($n_i$)

Set the fatigue life of the detail as X years.
The number of cycles for each stress range $S_i$ can be expressed in terms of X and the stress cycles per day ($S_{i,pd}$), i.e.

$$n_i = 365 \cdot X \cdot n_{i,pd}$$

3. Determine fatigue life

From the fatigue damage accumulation D < 1, i.e.

$$D = \sum_{i=1}^{3} \frac{n_i}{N_i} = \frac{n_1}{N_1} + \frac{n_2}{N_2} + \frac{n_3}{N_3} = 365 \cdot X \cdot \left( \frac{n_{1,pd}}{N_1} + \frac{n_{2,pd}}{N_2} + \frac{n_{3,pd}}{N_3} \right) < 1$$

$$X < \frac{1}{365 \cdot \left( \frac{n_{1,pd}}{N_1} + \frac{n_{2,pd}}{N_2} + \frac{n_{3,pd}}{N_3} \right)} = \frac{1}{365 \cdot \left( \frac{30}{0.48 \cdot 10^6} + \frac{100}{1.14 \cdot 10^6} + \frac{1000}{6.08 \cdot 10^6} \right)} \approx 9 \text{ years}$$

## Solution according to Eurocode 3 Part 1.9

The solution according to Eurocode 3 Part 1.9 is the same as that given in AS 4100 except that a partial safety factor ($\gamma_{Mf}$) is used instead of a capacity factor.

From the solution of Section 8.5.3 Example 3

$$N_1 = 0.57 \cdot 10^6$$
$$N_2 = 1.35 \cdot 10^6$$
$$N_3 = 8.07 \cdot 10^6$$

Similarly

$$X < \frac{1}{365 \cdot \left( \frac{n_{1,pd}}{N_1} + \frac{n_{2,pd}}{N_2} + \frac{n_{3,pd}}{N_3} \right)} = \frac{1}{365 \cdot \left( \frac{30}{0.57 \cdot 10^6} + \frac{100}{1.34 \cdot 10^6} + \frac{1000}{8.07 \cdot 10^6} \right)} \approx 11 \text{ years}$$

## Solution according to BS 7608

1. Determine the fatigue strength $N_i$ at the stress range of $S_i$

From the solution of Section 8.5.3 Example 3

$$N_1 = 0.50 \cdot 10^6$$
$$N_2 = 1.18 \cdot 10^6$$
$$N_3 = 5.96 \cdot 10^6$$
$$N_4 = 71.4 \cdot 10^6$$

2. Determine fatigue life

From the fatigue damage accumulation $D < 1$, i.e.

$$D = \sum_{i=1}^{4} \frac{n_i}{N_i} = \frac{n_1}{N_1} + \frac{n_2}{N_2} + \frac{n_3}{N_3} + \frac{n_4}{N_4} = 365 \cdot X \cdot \left( \frac{n_{1,pd}}{N_1} + \frac{n_{2,pd}}{N_2} + \frac{n_{3,pd}}{N_3} + \frac{n_{4,pd}}{N_4} \right) < 1$$

$$X < \frac{1}{365 \cdot \left( \frac{n_{1,pd}}{N_1} + \frac{n_{2,pd}}{N_2} + \frac{n_{3,pd}}{N_3} + \frac{n_{4,pd}}{N_4} \right)}$$

$$= \frac{1}{365 \cdot \left( \frac{30}{0.50 \cdot 10^6} + \frac{100}{1.18 \cdot 10^6} + \frac{1000}{5.96 \cdot 10^6} + \frac{2000}{71.4 \cdot 10^6} \right)} \approx 8 \text{ years}$$

Note: The predicted fatigue life using AS 4100, Eurocode 3 Part 1.9 and BS 7608 is very close ranging from 8 to 11 years. A larger fatigue life would be obtained if a smaller partial safety factor $\gamma_{Mf}$ was adopted when using BS 7608. It can be proven that the fatigue life would become 24 years and 13 years if the partial safety factor $\gamma_{Mf}$ was taken as 1.0 and 1.15 respectively.

### 8.5.5 *Example 5*

Determine the fatigue life of the detail described in Section 8.5.4 Example 4 if the capacity factor $\phi$ = 1.0 and the partial safety factor $\gamma_{Mf}$ = 1.0 (i.e. the necessary conditions specified in Section 8.2.2 are satisfied, e.g. the detail is located in a redundant load path and is accessible for, and subject to, regular inspection).

**Solution according to AS 4100**

1. Determine the fatigue strength $N_i$ at the stress range of $S_i$

$\phi S_{CA} = 1.0 \times 52 = 52$ N/mm$^2$
$\phi S_{CO} = 1.0 \times 29 = 29$ N/mm$^2$

Since $S_1$ and $S_2$ are larger than $\phi S_{CA}$, use Equation (8.3a)

$$N_1 = \frac{(\phi \cdot S_{DC})^3 \cdot 2 \cdot 10^6}{S_1^3} = \frac{(1.0 \times 71)^3 \cdot 2 \cdot 10^6}{80^3} = 1.40 \cdot 10^6$$

$$N_2 = \frac{(\phi \cdot S_{DC})^3 \cdot 2 \cdot 10^6}{S_2^3} = \frac{(1.0 \times 71)^3 \cdot 2 \cdot 10^6}{60^3} = 3.32 \cdot 10^6$$

Since $\phi S_{CO} \leq S_3 < \phi S_{CA}$, use Equation (8.3b)

$$N_3 = \frac{(\phi \cdot S_{CA})^5 \cdot 5 \cdot 10^6}{S_3^5} = \frac{(1.0 \times 52)^5 \cdot 5 \cdot 10^6}{35^5} = 36.2 \cdot 10^6$$

2. Determine fatigue life

$$X < \frac{1}{365 \cdot \left( \frac{n_{1,pd}}{N_1} + \frac{n_{2,pd}}{N_2} + \frac{n_{3,pd}}{N_3} \right)} = \frac{1}{365 \cdot \left( \frac{30}{1.40 \cdot 10^6} + \frac{100}{3.32 \cdot 10^6} + \frac{1000}{36.2 \cdot 10^6} \right)} \approx 35 \text{ years}$$

**Solution according to Eurocode 3 Part 1.9**

The solution according to Eurocode 3 Part 1.9 is the same as that given in AS 4100 because $\gamma_{Mf} = \phi = 1.0$ in this case, i.e. the fatigue life is about 35 years.

**Solution according to BS 7608**

1. Determine the fatigue strength $N_i$ at the stress range of $S_i$

$S_{CA}/\gamma_{Mf} = 40/1.0 = 40\ \text{N/mm}^2$

Since $S_1$ and $S_2$ are larger than $S_{CA}/\gamma_{Mf}$, use Equation (8.6a)

$$N_1 = \frac{(S_{DC}/\gamma_{Mf})^3 \cdot 2 \cdot 10^6}{S_1^3} = \frac{(68/1.0)^3 \cdot 2 \cdot 10^6}{80^3} = 1.23 \cdot 10^6$$

$$N_2 = \frac{(S_{DC}/\gamma_{Mf})^3 \cdot 2 \cdot 10^6}{S_2^3} = \frac{(68/1.0)^3 \cdot 2 \cdot 10^6}{60^3} = 2.90 \cdot 10^6$$

Since $S_3$ and $S_4$ are less than $S_{CA}/\gamma_{Mf}$, use Equation (8.6b)

$$N_3 = \frac{(S_{CA}/\gamma_{Mf})^5 \cdot 10^7}{S_3^5} = \frac{(40/1.0)^5 \cdot 10^7}{35^5} = 19.5 \cdot 10^6$$

$$N_4 = \frac{(S_{CA}/\gamma_{Mf})^5 \cdot 10^7}{S_4^5} = \frac{(40/1.0)^5 \cdot 10^7}{20^5} = 320 \cdot 10^6$$

2. Determine fatigue life

$$X < \frac{1}{365 \cdot \left( \frac{n_{1,pd}}{N_1} + \frac{n_{2,pd}}{N_2} + \frac{n_{3,pd}}{N_3} + \frac{n_{4,pd}}{N_4} \right)}$$

$$= \frac{1}{365 \cdot \left( \frac{30}{1.23 \cdot 10^6} + \frac{100}{2.90 \cdot 10^6} + \frac{1000}{19.5 \cdot 10^6} + \frac{2000}{320 \cdot 10^6} \right)} \approx 24 \text{ years}$$

Note: The predicted fatigue life varies from 24 to 35 years that are approximately 3 to 4 times those given in Section 8.5.4 Example 4 with $\phi = 0.7$ and $\gamma_{Mf} = 1.35$.

## 8.6 References


1. BSI (1993), Fatigue design and assessment of steel structures, BS 7608, British Standard Institution, London, UK
2. BSI (1988), Steel, concrete and composite bridges, BS 5400, British Standard Institution, London, UK
3. BSI (2000), Structural use of Steelwork in Building, BS 5950, Part 1, British Standard Institution, London, UK
4. EC3 (2003), Eurocode 3: Design of Steel Structures – Part 1.9: Fatigue, prEN 1993-1-9: 2003, November 2003, European Committee for Standardization, Brussels, Belgium
5. EC1 (2002), Eurocode 1: Actions on Structures – Part 2: Traffic Loads on Bridges, prEN 1991-2, 10 January 2002, European Committee for Standardization, Brussels, Belgium
6. Fisher, J., Frank, K.H., Hirt, M.A. and McNamee, B.M. (1970), Effect of Weldment on Fatigue Strength of Steel Beam, *NCHRP Report No. 102*, Lehigh University, Bethlehem, Pennsylvania, USA
7. Grundy, P. (2004), Fatigue Design of Steel Structures, *Steel Construction*, Journal of the Australian Steel Institute, 38(1), pp 1-8
8. Gurney, T.R. (1979), *Fatigue of Welded Structures*, Cambridge University Press, 2nd Edition, Cambridge, UK
9. Kurobane, Y. (1989), Recent Development in the Fatigue Design Rules in Japan, *Proceedings*, International Symposium on the Occasion of the Retirement of Prof. J. de Back, Delft, The Netherlands, pp 173-187
10. Maddox, S. J. (1991), *Fatigue Strength of Welded Structures*, Abington Publishing, Cambridge, UK
11. Mang, F. and Bucak, Ö (1982), Fatigue Behaviour of Welded Joints in Trusses of Steel Hollow Sections, *Proceedings*, IABSE Colloquium on Fatigue of Steel and Concrete Structures, Lausanne, Switzerland, pp 735-744
12. Marshall, P.W. (1992), *Design of Welded Tubular Connections – Basis and Use of AWS Code Provisions*, Elsevier Science Publishers, Amsterdam, The Netherlands
13. Mashiri, F.R., Zhao, X.L. and Grundy, P. (2002), Fatigue Tests and Design of Welded T Connections in Thin Cold-Formed Square Hollow Sections under In-Plane Bending, *Journal of Structural Engineering*, ASCE, 128 (11), pp 1413-1422
14. Mashiri, F.R., Zhao, X.L. and Grundy, P. (2004), Fatigue of Cold-Formed CHS Joints ($t < 4$ mm), *CIDECT Report No. 7U-9/04*, Department of Civil Engineering, Monash University, Melbourne, Australia
15. Niemi, E. (1999), Designer's Guide for Hot Spot Fatigue Analysis, *IIW Doc. XIII-WG3-06-99*, IIW Annual Assembly, Lisbon, Portugal
16. Standards Australia (1998), Steel Structures, Australian Standard AS 4100, Standards Australia, Sydney, Australia
17. Standards Australia (1999), Supplement to AS 4100-1998, Standards Australia, Sydney, Australia
18. Standards Australia (2002), Cranes, Hoists and Winches, Australian Standard AS1418, Standards Australia, Sydney, Australia
19. Zhao, X.L., Sedlacek, G. and Schäfers, M. (1999), Fatigue Resistance of Welded Tubular Joints using Fracture Mechanics Approach, *Proceedings*, 2nd Australasian Congress on Applied Mechanics, February, Canberra, CD-Rom, Paper R-064

20. Zhao, X.L., Herion, S., Packer, J.A., Puthli, R., Sedlacek, G., Wardenier, J., Weynand, K., van Wingerde, A. and Yeomans, N. (2000), *Design Guide for Circular and Rectangular Hollow Section Welded Joints under Fatigue Loading*, TÜV- Verlag, Köln, Germany

# Chapter 9: Recent Developments

This chapter presents some recent developments in the design of cold-formed tubular members and connections which are not covered in AS 4100 (Standards Australia 1998) and BS 5950 Part 1 (BSI 2000). The issues discussed are limiting width-to-thickness ratios for concrete-filled tubes and for those subjected to large-deformation cyclic loading, fatigue design using the hot spot stress method, bolted moment end plate connections and plastic design of portal frames. Other recent developments are also briefly mentioned in Section 9.5.

## 9.1 Effect of Concrete-Filling and Large Deformation Cyclic Loading on Limiting Width-to-Thickness Ratios

### *9.1.1 General*

As described in Chapter 3 different classification systems of steel sections exist in various design codes. Eurocode 3 Part 1.1 (EC 3 2003) classifies sections as Class 1, 2, 3 or 4. BS 5950 Part 1 classifies sections as plastic, compact, semi-compact and slender. AS 4100 and AISC LRFD (1999) classify sections as compact, non-compact and slender. The category to which a section belongs depends on its cross-section geometry and certain limits on such geometry specified in the design code. The concept of plate element slenderness is used in AS 4100 whereas the actual width-to-thickness ratio or diameter-to-thickness ratio is used in Eurocode 3 Part 1.1, BS 5950 Part 1 and the AISC LRFD specification. The limiting width-to-thickness ratios are given in Chapters 3 and 4.

It is well known that in general concrete-filling increases the limiting width-to-thickness ratio whereas the large-deformation cyclic loading decreases the limiting ratio (Bergmann et al 1995). In this chapter, the limits for unfilled tubes under static loading are compared with the limits for concrete-filled tubes under static loading, limits for unfilled tubes under cyclic loading and those for concrete-filled tubes under cyclic loading. The limits for unfilled tubes under static loading specified in Eurocode 3 Part 1.1 are chosen as the reference value for comparison. This is because the major standards (Eurocode 4 Part 1.1 2001, Eurocode 8 Part 1.1 2003 and AISC 2002), which give the other three limits, have the same definition of width-to-thickness ratios using the concept of flat width defined in Figure 9.1 where symbols in Eurocode 3 Part 1.1 are adopted. In Figure 9.1 r is the internal corner radius.

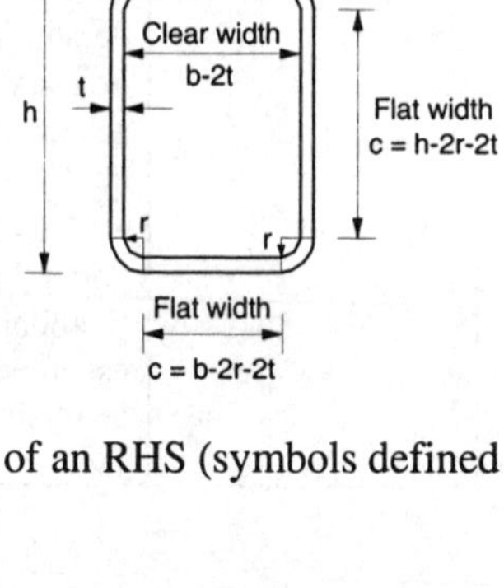


Figure 9.1 Geometry of an RHS (symbols defined in Eurocode 3 Part 1.1)

The limiting width-to-thickness ratios specified in Eurocode 3 Part 1.1 are summarised in Table 9.1 where c is the flat width of an RHS defined in Figure 9.1, d is the outside diameter of a CHS, t is the tube wall thickness, subscript 1 refers to Class 1 section, subscript 3 refers to Class 3 section and $f_y$ is the yield stress in $N/mm^2$.

Table 9.1 Limiting width-to-thickness ratios for unfilled tubes under static loading

| Case | RHS bending (Flange) | CHS bending | RHS compression | CHS compression |
|---|---|---|---|---|
| Limiting width-to-thickness ratio | $(\frac{c}{t})_{1,EC3}$ | $(\frac{d}{t})_{1,EC3}$ | $(\frac{c}{t})_{3,EC3}$ | $(\frac{d}{t})_{3,EC3}$ |
| Value | $33 \cdot \sqrt{\frac{235}{f_y}}$ | $50 \cdot \frac{235}{f_y}$ | $42 \cdot \sqrt{\frac{235}{f_y}}$ | $90 \cdot \frac{235}{f_y}$ |

### 9.1.2 *Effect of Concrete-Filling*

The effect of concrete-filling on the limiting width-to-thickness ratios under static loading can be demonstrated by the comparison shown in Table 9.2. The increase in the limiting values ranges from 20% to 80%. However, the same limiting value is used in Eurocode 4 Part 1.1 (2001) for concrete-filled CHS in compression as those for unfilled CHS.

Table 9.2 Effect of concrete-filling on the limiting width-to-thickness ratios under static loading

| Case | RHS bending (Flange) | CHS bending | RHS compression | CHS compression |
|---|---|---|---|---|
| Ratios | $\frac{(\frac{c}{t})_{1,filled,static}}{(\frac{c}{t})_{1,EC3}}$ | $\frac{(\frac{d}{t})_{1,filled,static}}{(\frac{d}{t})_{1,EC3}}$ | $\frac{(\frac{c}{t})_{3,filled,static}}{(\frac{c}{t})_{3,EC3}}$ | $\frac{(\frac{d}{t})_{3,filled,static}}{(\frac{d}{t})_{3,EC3}}$ |
| Values (Reference) | ≈1.6 (EC4 Part 1.1 2001) | = 1.8 (EC 4 Part 1.1 2001) | ≈1.2 (EC4 Part 1.1 2001) | = 1.0 (EC4 Part 1.1 2001) |
| | | ≈1.5 (Elchalakani et al 2001) | ≈1.5 (Matsui et al 1997) | |
| | | | ≈1.7 (AIJ 1997) | ≈1.7 (AIJ 1997) |
| | | | ≈1.2 to 1.5 depending on the boundary conditions assumed in the analysis (Uy 2000) | |
| | | | ≈1.2 to 1.7 depending on the boundary conditions assumed in the analysis (Wright 1995) | |

### 9.1.3 *Effect of Large-Deformation Cyclic Loading*

The effect of large-deformation cyclic loading on the limiting width-to-thickness ratios for unfilled tubes can be illustrated by the comparisons shown in Table 9.3. The limiting ratios range from about 40% to 80% of those under static loading. The loading scheme and ductility requirement play an important role in determining the limiting width-to-thickness ratios under cyclic loading.

Table 9.3 Effect of large deformation cyclic loading on the limiting width-to-thickness ratios for unfilled tubes

| Case | RHS bending (Flange) | CHS bending | RHS compression | CHS compression |
|---|---|---|---|---|
| Ratios | $\frac{(\frac{c}{t})_{1,cyclic}}{(\frac{c}{t})_{1,EC3}}$ | $\frac{(\frac{d}{t})_{1,cyclic}}{(\frac{d}{t})_{1,EC3}}$ | $\frac{(\frac{c}{t})_{3,cyclic}}{(\frac{c}{t})_{3,EC3}}$ | $\frac{(\frac{d}{t})_{3,cyclic}}{(\frac{d}{t})_{3,EC3}}$ |
| Values (Reference) | ≈ 0.6 to 0.8 (based on test data in Zhao and Hancock (1991) with a required rotational capacity of 7 to 9 specified in AISC (2002) for cyclic loading) | ≈ 0.72 (Elchalakani et al 2004a) | ≈ 0.46 (AISC 2002) | ≈ 0.42 (AISC 2002) |
| | | | | ≈ 0.6 (NZS 3404 1992) |
| | | | | ≈ 0.6 to 0.8 (Elchalakani et al 2003) |

### 9.1.4 *Combined Effect of Concrete-Filling and Large-Deformation Cyclic Loading*

The combined effect of concrete-filling and large-deformation cyclic loading on the limiting width-to-thickness ratios can be illustrated by the comparison shown in Table 9.4. It is interesting to observe a mixed effect with a ratio ranging from 0.6 to 1.6. Again the loading scheme and ductility requirement play an important role in determining the limiting width-to-thickness ratios under cyclic loading (Elchalakani et al 2004b, Elchalakani 2003).

Table 9.4 Effect of concrete-filling and large deformation cyclic loading on the limiting width-to-thickness ratios

| Case | RHS bending (Flange) | CHS bending | RHS compression | CHS compression |
|---|---|---|---|---|
| Ratios | $\frac{(\frac{c}{t})_{1,filled,cyclic}}{(\frac{c}{t})_{1,EC3}}$ | $\frac{(\frac{d}{t})_{1,filled,cyclic}}{(\frac{d}{t})_{1,EC3}}$ | $\frac{(\frac{c}{t})_{3,filled,cyclic}}{(\frac{c}{t})_{3,EC3}}$ | $\frac{(\frac{d}{t})_{3,filled,cyclic}}{(\frac{d}{t})_{3,EC3}}$ |
| Values (Reference) | ≈ 0.7 to 1.5 (EC8 2003) | ≈ 1.6 (EC8 2003) | ≈ 0.6 to 1.2 (EC8 2003) | ≈ 0.9 (EC8 2003) |
| | | ≈ 0.7 to 1.0 (Elchalakani 2003, Elchalakani et al 2004b) | ≈ 1.0 (AISC 2002) | |

### *9.1.5 Summary*

Based on the above discussions, a few very simple rules may be summarised as follows:

1. Concrete-filling increases the limiting width-to-thickness ratios by approximately 50%.
2. Large deformation cyclic loading reduces the limiting width-to-thickness ratios by approximately 50%.
3. The limiting width-to-thickness ratios may increase or reduce by approximately 50% due to the combined effect of concrete-filling and large deformation cyclic loading depending on the ductility requirement and cyclic loading schemes. If the ductility requirement is low and the loading scheme is not severe, the limiting width-to-thickness ratios are mainly influenced by the concrete-filling, i.e. they may increase up to 50%. If the ductility requirement is high and the loading scheme is severe, the limiting width-to-thickness ratios are mainly influenced by the cyclic loading, i.e. they may reduce up to 50%.

## 9.2 Fatigue Design using the Hot Spot Stress Method

### *9.2.1 General*

In welded tubular joints the stiffness around the intersection is not uniform, resulting in a geometrical non-uniform stress distribution. The hot spot stress (also called geometric stress) method relates the fatigue life of a joint to the hot spot stress at the joint rather than the nominal stress. It takes the uneven stress distribution around the perimeter of the joint directly into account. The hot spot stress range includes the influence of the geometry and type of load but excludes the effects related to fabrication such as the configuration of the weld (flat, convex, concave) and the local condition of the weld toe (radius of weld toe, undercut, etc). The hot spot stress is the maximum geometrical stress occurring in the joint where the cracks are usually initiated. The hot spot stress method was recommended by the International Institute of Welding (IIW) Subcommission XV-E (IIW 1985) for the design of welded tubular joints.

Design rules for welded CHS and RHS joints under fatigue loading have recently become available through the publication of the latest IIW Fatigue Design Procedure for Welded Hollow Section Joints (IIW 1999) and the CIDECT Design Guide No.8 (Zhao et al 2000). The types of joints covered consist of circular or square hollow sections as used in uniplanar or multiplanar trusses or girders, such as T, Y, X, K, XX and KK joints.

This section summarizes briefly the fatigue design procedures and SCF calculations presented in Zhao et al (2000). The same symbols used in Zhao et al (2000) are adopted here.

## 9.2.2 *Fatigue Design Procedures*

### *9.2.2.1 Design Procedures*

The fatigue design procedures can be summarised as follows:
Step 1 Determine the axial forces and bending moments in the chord and braces
Step 2 Determine the nominal stress ranges
Step 3 Determine the stress concentration factors (SCFs)
Step 4 Determine the hot spot stress ranges
Step 5 Determine the permissible number of cycles for a given hot spot stress range at a specific joint location from a fatigue strength curve
Step 6 Determine the fatigue damage accumulation

### *9.2.2.2. Member Forces*

For welded hollow section structures, member forces must be obtained by analysis of the complete structure, in which nodal eccentricity of the member centrelines at the joint (connection) as well as local joint flexibility are taken into account. This can be achieved by either:
(a) a sophisticated three dimensional finite element modelling where plate, shell and solid elements are used at the joints, or
(b) a rigid frame analysis for two- or three-dimensional Vierendeel girders, or
(c) a simplified structural analysis using plane frame analysis for triangulated trusses or lattice girders. Axial forces and bending moments in the members can be determined using a structural analysis assuming a continuous chord and pin-ended braces. This produces axial forces in the braces, and both axial forces and bending moments in the chord. This modelling assumption is particularly appropriate for moving loads along the chord members in structures such as cranes and bridges.

### *9.2.2.3 Nominal Stress Ranges*

The determination of nominal stress ranges depends on the method used to determine member forces.

If analysis has been undertaken using the sophisticated three-dimensional finite element modelling or using the rigid frame analysis for two- or three-dimensional Vierendeel girders, the nominal stress range in any member can be determined using elastic beam theory.

If analysis has been undertaken using the plane frame analysis for triangulated trusses or lattice girders, assuming a continuous chord and pin-ended braces, the nominal stress range in any member can be determined using elastic beam theory except that a magnification factor (MF) should be applied to the axial stress ranges to account for secondary bending moments in K-joints. The value of the magnification factor varies from 1.3 to 1.8 as listed in Table 8.9 of Chapter 8.

#### *9.2.2.4 Stress Concentration Factor*

The stress concentration factor (SCF) is the ratio between the hot spot stress at the joint and the nominal stress in the member due to a basic member load which causes this hot spot stress. Formulae and graphs for calculating SCFs are presented in Zhao et al (2000). Some remarks are given later in Section 9.2.3.

#### *9.2.2.5 Hot Spot Stresses*

If the analysis has been undertaken using sophisticated three-dimensional finite element modelling, the hot spot stress ranges may be directly obtained from the analysis for each load combination. In all other cases the following procedures should be used to determine the hot spot stress ranges.

- The hot spot stress range at one location under one load case is the product of the nominal stress range and the corresponding stress concentration factor (SCF).
- Superposition of the hot spot stress ranges at the same location can be used for combined load cases.
- If the position of the maximum hot spot stress in a member, for the relevant loading condition, cannot be determined, then the maximum SCF values must be applied to all points around the periphery of the member at a joint.
- Hot spot stress ranges must be calculated for both the chord member and brace members.

Examples can be found in Zhao et al (2000).

#### *9.2.2.6 Fatigue Strength Curves*

Similar to the classification method, fatigue strength curves ($S_{rhs} - N_f$ curves) are used in the design, where $S_{rhs}$ is the hot spot stress ranges. A basic $S_{rhs} - N_f$ curve is used for hollow section joints with a wall thickness of 16 mm (Thorpe and Sharp 1989, DEn 1993, Dimitrakis et al 1995, van Wingerde 1996, 1997a, 1997b). For joints with wall thickness other than 16 mm, thickness correction factors are introduced. The influence of the thickness effect on fatigue behaviour of hollow section joints has been investigated by Gurney (1979), van Delft (1981), Marshall (1984, 1992), van Delft et al (1985), Berge and Webster (1987), Haagensen (1989), Thorpe and Sharp (1989), van Wingerde (1992). The thickness effect usually results in higher $S_{rhs} - N_f$ curves for smaller wall thickness. A common set of $S_{rhs} - N_f$ curves and thickness correction formulae for both CHS and RHS joints were presented in van Wingerde et al (1997c, 1998a). They were adopted by IIW (1999) and Zhao et al (2000), as shown in Figure 9.2.

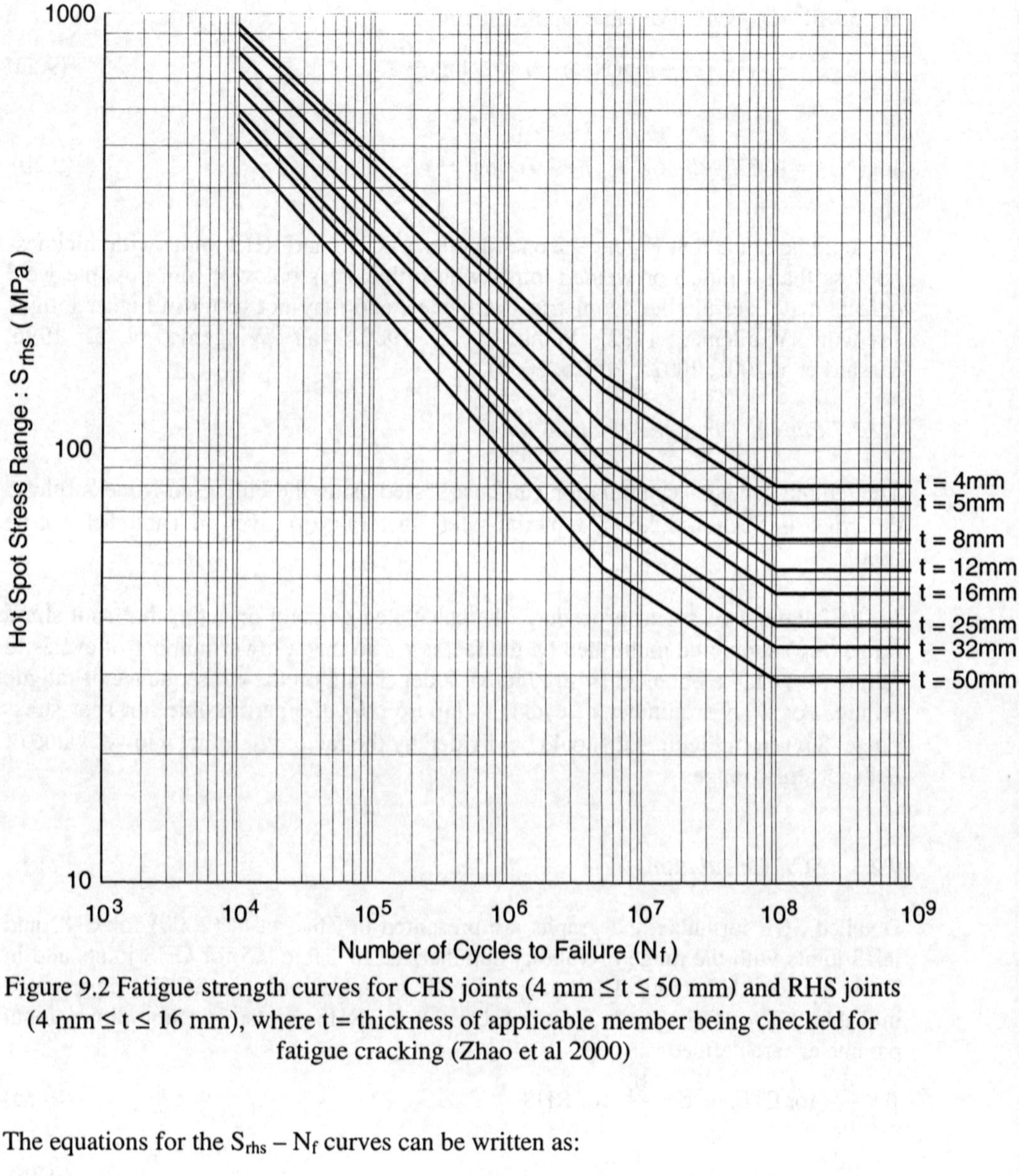


Figure 9.2 Fatigue strength curves for CHS joints (4 mm ≤ t ≤ 50 mm) and RHS joints (4 mm ≤ t ≤ 16 mm), where t = thickness of applicable member being checked for fatigue cracking (Zhao et al 2000)

The equations for the $S_{rhs} - N_f$ curves can be written as:

For $10^3 < N_f < 5 \cdot 10^6$

$$\log(S_{rhs}) = \frac{1}{3} \cdot (12.476 - \log(N_f)) + 0.06 \cdot \log(N_f) \cdot \log(\frac{16}{t}) \quad (9.1a)$$

or

$$\log(N_f) = \frac{12.476 - 3 \cdot \log(S_{rhs})}{1 - 0.18 \cdot \log(\frac{16}{t})} \quad (9.1b)$$

For $5 \cdot 10^6 < N_f < 10^8$ (variable amplitude only)

$$\log(S_{rhs}) = \frac{1}{5} \cdot (16.327 - \log(N_f)) + 0.402 \cdot \log(\frac{16}{t}) \quad (9.2a)$$

or

$$\log(N_f) = 16.327 - 5 \cdot \log(S_{rhs}) + 2.01 \cdot \log(\frac{16}{t}) \quad (9.2b)$$

It should be noted that Figure 9.2 only applies to CHS and RHS joints with thickness not less than 4 mm. For welded joints with a thickness below 4 mm possible weld defects may overrule the geometrical influence, and may not result in higher fatigue strength (Wardenier 1982, Puthli et al 1989, van Wingerde et al 1996, Mashiri et al 2001, 2002a, 2002b, 2004).

### 9.2.2.7 *Fatigue Damage Accumulation*

The fatigue damage accumulation can be assessed using the Palmgren-Miner's rule as described in Section 8.2.6. The partial safety factors ($\gamma_{Mf}$) given in Table 8.1 can be used.

In the fatigue assessment procedure, the calculated demand or acting hot spot stress range ($S_{rhs}$) should be multiplied by the factor $\gamma_{Mf}$ so that a lower number of cycles to failure ($N_f$) can be obtained from Figure 9.2 depending on the consequence of fatigue failure. For a given number of cycles to failure ($N_f$), the permissible hot spot stress range ($S_{rhs}$) from Figure 9.2 should be divided by the factor $\gamma_{Mf}$ to get a lower value of hot spot stress range.

## 9.2.3 *SCF Calculations*

Detailed SCF formulae and graphs are presented in Zhao et al (2000) for CHS and RHS joints with the range of validity summarised in Table 9.5 for CHS joints and in Table 9.6 for RHS joints. The hot spot locations and remarks on SCFs are presented in Table 9.7 for CHS joints and in Table 9.8 for RHS joints. The non-dimensional parameters are defined as:

$$\beta = \frac{d_1}{d_0} \text{ for CHS or } \beta = \frac{b_1}{b_0} \text{ for RHS} \quad (9.3a)$$

$$\gamma = \frac{d_0}{2 \cdot t_0} \text{ for CHS or } \gamma = \frac{b_0}{2 \cdot t_0} \text{ for RHS} \quad (9.3b)$$

$$\tau = \frac{t_1}{t_0} \quad (9.3c)$$

$$\alpha = \frac{2 \cdot L}{d_0} \quad (9.3d)$$

The other symbols used in this section are also defined in the diagrams in Table 9.6, Table 9.7 and Table 9.8.

Table 9.5 Range of validity for SCF formulae and graphs (CHS joints)

| T, Y and X-joints | K-joints with gap | XX-joints | KK-joints with gap |
|---|---|---|---|
| $0.2 \le \beta \le 1.0$<br>$15 \le 2\gamma \le 64$<br>$0.2 \le \tau \le 1.0$<br>$4 \le \alpha \le 40$<br>$30° \le \theta \le 90°$ | No eccentricity<br>Equal braces<br>$0.3 \le \beta \le 0.6$<br>$24 \le 2\gamma \le 60$<br>$0.25 \le \tau \le 1.0$<br>$30° \le \theta \le 60°$ | No eccentricity<br>Equal braces<br>$0.3 \le \beta \le 0.6$<br>$15 \le 2\gamma \le 64$<br>$0.25 \le \tau \le 1.0$<br>$\theta = 90°$<br>$\phi = 90°$<br>$\psi = \phi - 2 \cdot \arcsin(\beta)$<br>$\ge 16.2°$ | No eccentricity<br>Equal braces<br>$0.3 \le \beta \le \cos(\theta)$<br>$24 \le 2\gamma \le 48$<br>$0.25 \le \tau \le 1.0$<br>$30° \le \theta \le 60°$<br>$60° \le \phi \le 180°$ |

Table 9.6 Range of validity for SCF formulae and graphs (RHS joints)

| T and X-joints | K-joints with gap | K-joints with overlap | KK-joints with gap |
|---|---|---|---|
| $0.35 \le \beta \le 1.0$<br>$12.5 \le 2\gamma \le 25.0$<br>$0.25 \le \tau \le 1.0$ | Equal braces<br>$0.35 \le \beta \le 1.0$<br>$10 \le 2\gamma \le 35$<br>$0.25 \le \tau \le 1.0$<br>$30° \le \theta \le 60°$<br>$2\tau \le g/t_o$<br>$-0.55 \le e/h_o \le 0.25$ | Equal braces<br>$0.35 \le \beta \le 1.0$<br>$10 \le 2\gamma \le 35$<br>$0.25 \le \tau \le 1.0$<br>$30° \le \theta \le 60°$<br>$50\% \le O_v \le 100\%$<br>$-0.55 \le e/h_o \le 0.25$<br>overlap<br>$O_v = (q/p) \times 100\%$<br>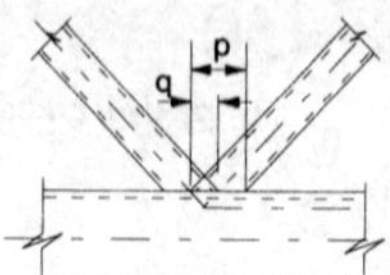 | Equal braces<br>$0.25 \le \beta \le 0.60$<br>$12.5 \le 2\gamma \le 25.0$<br>$0.5 \le \tau \le 1.0$<br>$30° \le \theta \le 60°$<br>$2\tau \le g/t_o$<br>$-0.55 \le e/h_o \le 0.25$<br>$60° \le \phi \le 180°$ |

Table 9.7 Hot spot locations and remarks on SCFs (CHS joints)

| Type of CHS joints | Remarks on SCFs |
|---|---|
| T, Y and X-joints<br>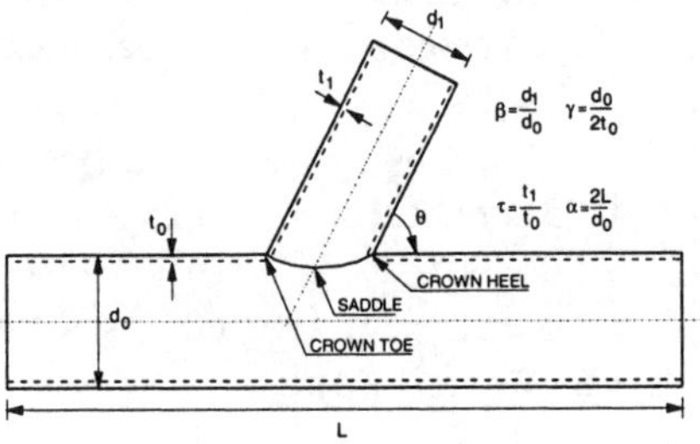<br> | • Generally the highest SCF occurs at the saddle position.<br>• Highest SCFs at the saddle are obtained for β ratios around 0.5 to 0.7.<br>• SCF decreases with decreasing τ value except for the brace crown position under axial loading.<br>• SCF decreases with decreasing 2γ value except for the brace crown position under axial loading.<br><br>References: Efthymiou and Durkin (1985), Efthymiou (1988) |
| K-joints with gap<br>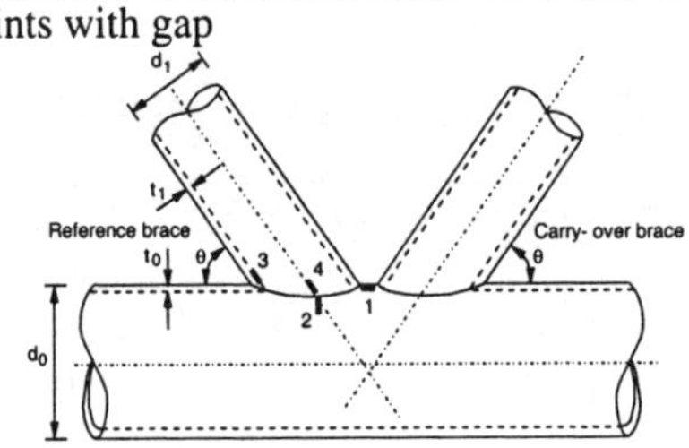<br> | • SCF decreases with decreasing τ value.<br>• SCF decreases with decreasing 2γ value.<br><br>References: Romeijn (1994), Dijkstra et al (1996), Karamanos et al (1997), van Wingerde et al (1998a) |
| XX-joints<br>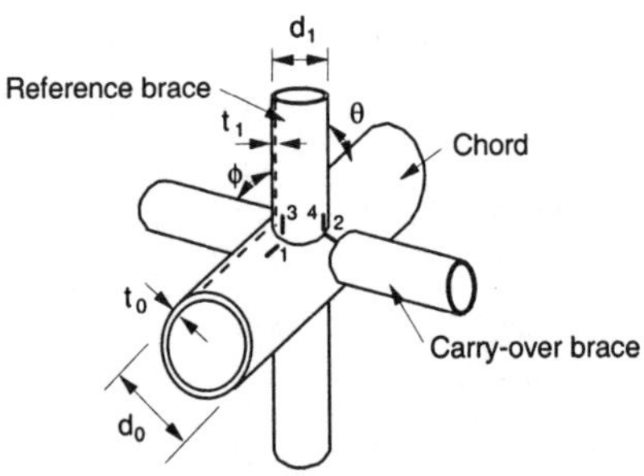<br> | Similar remarks as for CHS X-joints.<br><br>For multiplanar joints the load in one brace plane may affect the hot spot stress range in another brace plane. This is called the multiplanar effect or carry-over effect. This effect is considered only at the saddle locations for axial loading or out-of-plane bending. In-plane bending does not introduce any multiplanar phenomena. The chord loading effects are concentrated only on crown locations of the chord.<br><br>References: Romeijn (1994), Dijkstra et al (1996), Karamanos et al (1997) |
| KK-joints with gap<br>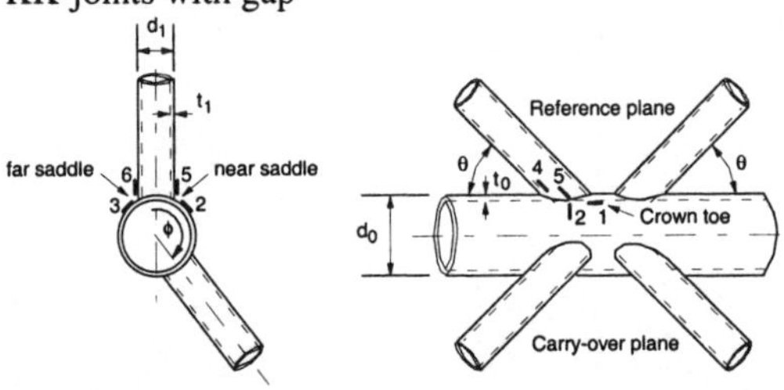<br> | Similar remarks as for CHS K-joints with gap.<br><br>For axial balanced brace loading, the multiplanar effects are only considered for anti-symmetrical load case.<br><br>References: Romeijn et al (1993), Romeijn (1994), Dijkstra et al (1996), Karamanos et al (1997) |

Table 9.8 Hot spot locations and remarks on SCFs (RHS joints)

<table>
<tr><th>Type of RHS joints</th><th>Remarks on SCFs</th></tr>
<tr><td>T and X-joints
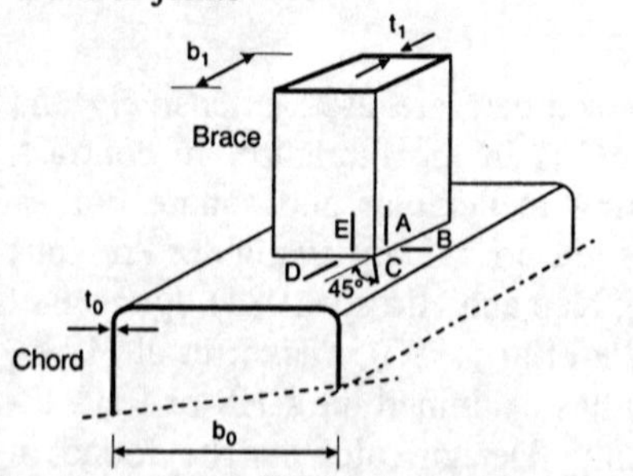

</td><td>

- The highest SCFs generally occur in the chord (for $\tau = 1$) at locations B and C.
- The highest SCFs are found for $\beta$ ratios around 0.5 to 0.7.
- The lower is the $2\gamma$ ratio, the lower is the SCF.
- The lower is the $\tau$ ratio, the lower is the SCF in the chord. $\tau$ has less influence on SCFs in the brace.

References: van Wingerde (1992)</td></tr>
<tr><td>K-joints with gap
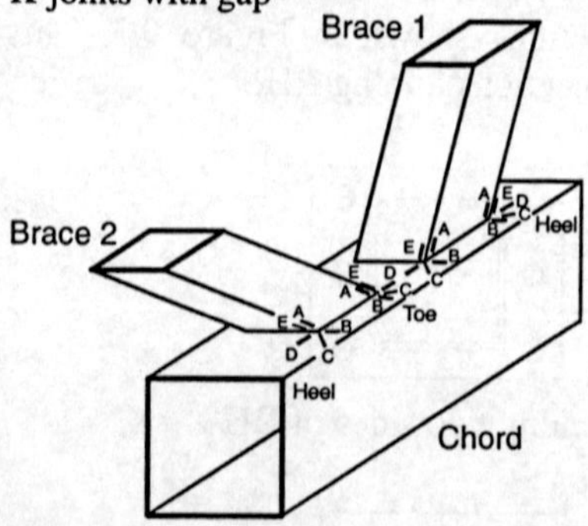

</td><td>

- Highest SCFs occur for $\beta$ ratios around 0.5 to 0.6.
- For the chord, SCF decreases as $2\gamma$ decreases, and SCF decreases as $\tau$ decreases.
- For the brace, SCF decreases as $2\gamma$ decreases, and SCF decreases as $\tau$ increases.

References: van Wingerde (1996, 1997a, 1997b), Puthli and Herion (1996), Zhao and Puthli (1998)</td></tr>
<tr><td>K-joints with overlap
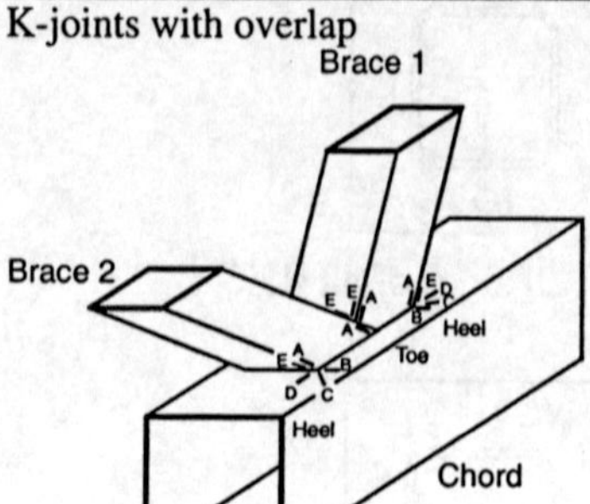

</td><td>

- Highest SCFs occur for RHS K-joints with 50% overlap and $\beta$ ratios around 0.5 to 0.7.
- For the chord, SCF decreases as $2\gamma$ decreases, and SCF decreases as $\tau$ decreases.
- For the brace, SCF decreases as $2\gamma$ decreases, and SCF decreases as $\tau$ increases.
- SCFs for overlapped K-joints are generally lower than those for gapped K-joints.

References: van Wingerde et al (1996, 1997a, 1997b)</td></tr>
<tr><td>KK-joints with gap
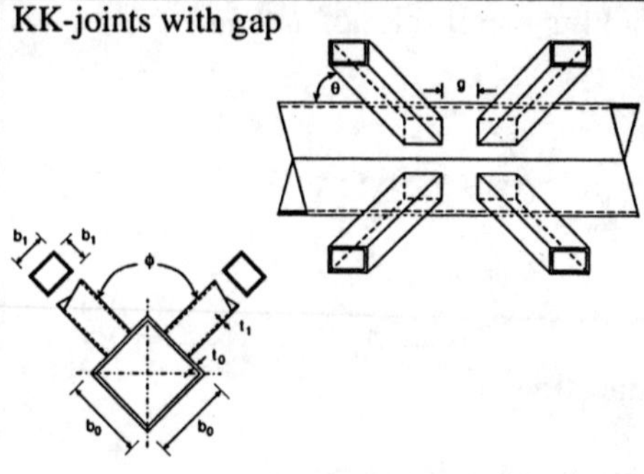

</td><td>Similar remarks as for RHS K-joints with gap.

For axial balanced brace loading, the multiplanar effects are only considered for anti-symmetrical load case.
For chord loading, no multiplanar correction is needed.

References: van Wingerde et al (1998b)</td></tr>
</table>

## 9.3 Bolted Moment End Plate Connections

### *9.3.1 General*

Moment end plate connections joining I-section members are used extensively and considerable documentation on their behaviour exists in the literature. In contrast, research on moment end plate connections joining rectangular and square hollow sections is limited and satisfactory design models are not widely available. Previous research by Mang (1980), Kato and Mukai (1982), Kao and Hirose (1984), Igarashi et al (1985), Birkemoe and Packer (1986), Watanabe et al (1997), Packer et al (1989, 1999), Cao et al (2000), Ranzi and Kneen (2002) has examined the CHS or RHS-to-flange connection mostly under axial loading only. Design rules can be found in CIDECT Design Guides No.1 (Wardenier et al 1991), No.3 (Packer et al 1992) and Packer and Henderson (1997) for bolted connections shown in Figure 9.3. This section addresses the bolted moment end plate connections joining RHS.

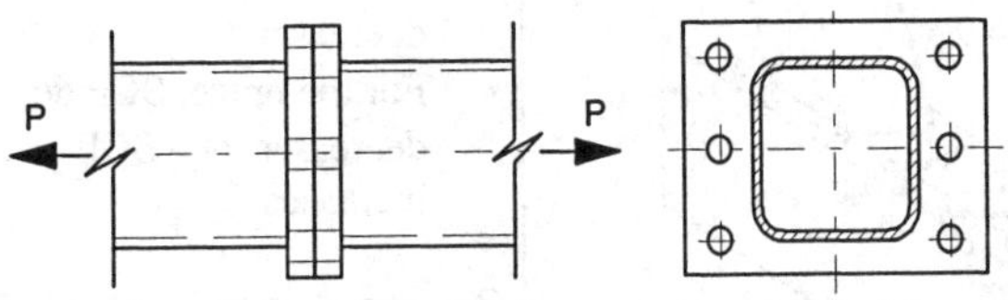


(a) Flange-plate connection with bolts along two sides of RHS

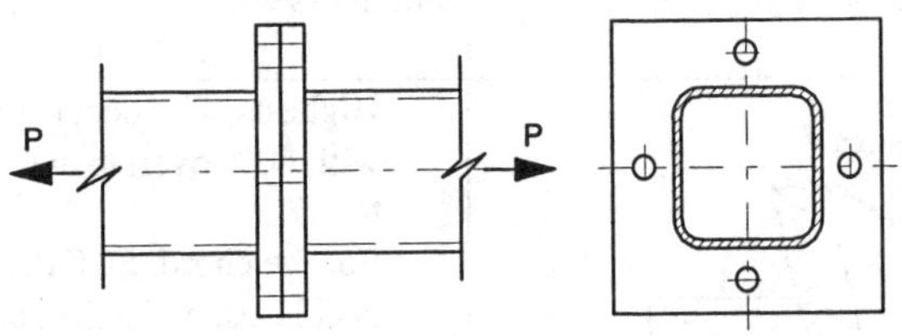


(b) Four bolt configurations for bolting on all sides of an RHS

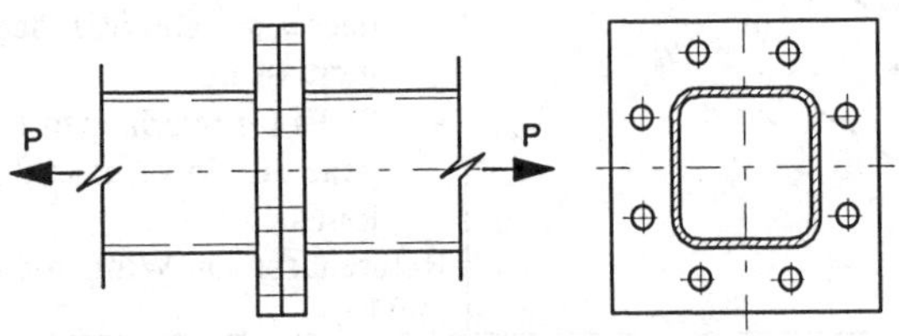


(c) Eight bolt configurations for bolting on all sides of an RHS

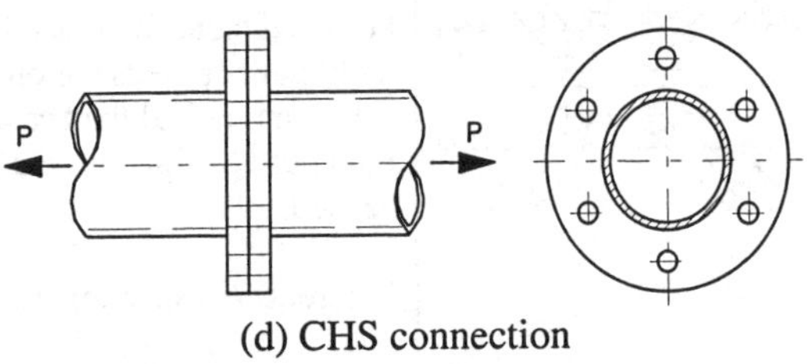


(d) CHS connection

Figure 9.3 Bolted connections under axial loading

### 9.3.2 Bolted Moment End Plate Behaviour

Tests were conducted by Wheeler et al (1995) on bolted moment end plate connections joining RHS. Figure 9.4 shows such a connection under testing. An analytical model to predict the serviceability limit moment and ultimate moment capacities of such connections has been presented in Wheeler et al (1998). The connection geometry considered utilizes two rows of bolts, one of which is located above the tension flange and the other of which is positioned symmetrically below the compression flange, as shown in Figure 9.5. Using a so-called modified stub-tee approach, the model considers the combined effects of prying action caused by flexible end plates and the formation of yield lines in the end plates as shown in Figure 9.6. The model has been calibrated against experimental data given in Wheeler et al (1995). Finite element analysis of such connections was also reported in Wheeler et al (2000).

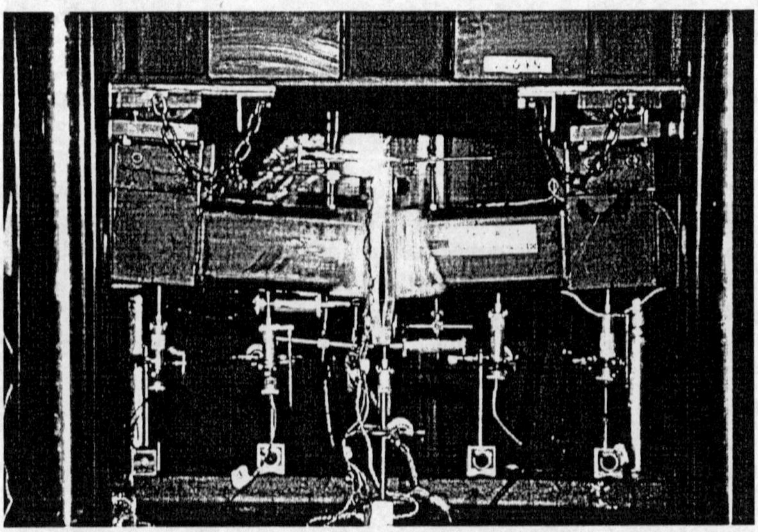

Figure 9.4 Bolted moment end plate connections under testing (Wheeler et al 1995)

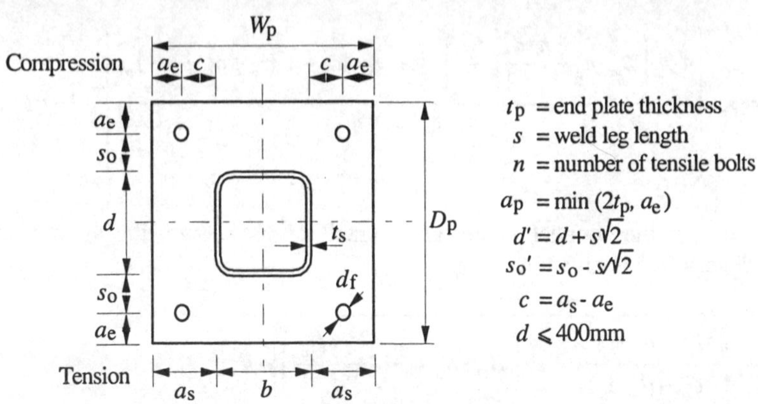


Figure 9.5 End plate layout and model parameters (Wheeler et al 1998)

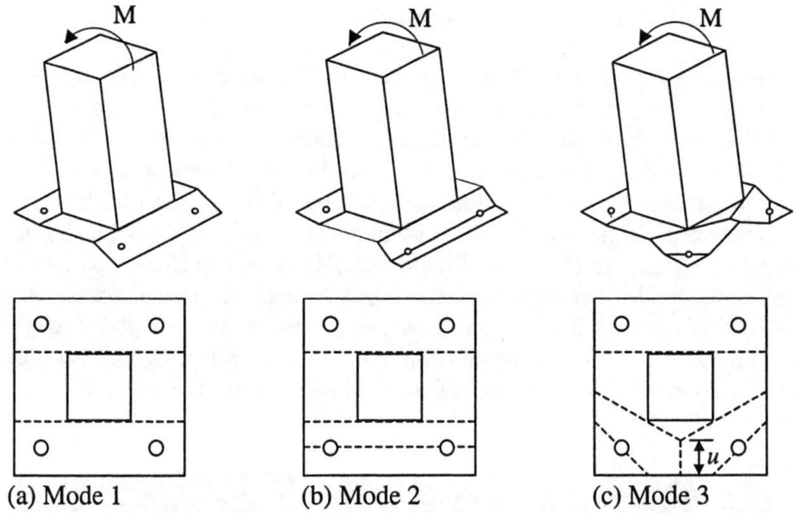


Figure 9.6 Yield line mechanisms for bolted moment end plate connection (Wheeler et al 1998)

Of the three types of end plate behaviour considered in the stub-tee model (thick, thin and intermediate), Wheeler et al (1998) recommended that the end plate connections be designed to behave in an intermediate fashion, with the connection strength being governed by tensile bolt failure. Thin plate behaviour results in connections that are more ductile and exhibit extremely high rotations, while connections exhibiting thick plate behaviour have much less rotation capacity and may be uneconomical.

### 9.3.3 *Connection Capacity*

Equations to calculate the connection capacity based on bolt failure and end plate failure presented in Wheeler et al (1998) are summarised below.

For strength limit state design based on bolt failure

$$\phi_b \cdot M_{cb} = \phi_b \cdot \left\{ \frac{4 \cdot n \cdot \left( B_{ul} \cdot a_p + \frac{\pi \cdot d_b^3 \cdot f_{yb}}{32} \right) \cdot d' + w_{eq} \cdot \left( d' + 2 \cdot (s_o' + a_p) \right) \cdot t_p^2 \cdot f_p}{4 \cdot (a_p + s_o') \cdot d'} \right\} \cdot (d - t_s) \qquad (9.4)$$

Alternatively, if the connection design moment $M^*$ is known, the appropriate end-plate thickness ($t_{bu}$) is given by

$$t_{bu} = \sqrt{\frac{4 \cdot \left\{ \frac{M^* \cdot (a_p + s_o')}{\phi_b \cdot (d - t_s)} - n \cdot \left( B_{ul} \cdot a_p + \frac{\pi \cdot d_b^3 \cdot f_{yb}}{32} \right) \right\} \cdot d'}{w_{eq} \cdot (d' + 2 \cdot (s_o' + a_p)) \cdot f_p}} \qquad (9.5)$$

To avoid thick plate behaviour, the limit on the plate thickness for capacity limited by bolt failure is:

$$t_p \le t_{max} = \sqrt{\frac{4 \cdot n \cdot B_{ul} \cdot s_o'}{w_{eq} \cdot f_p}} \quad (9.6)$$

For strength limit state design based on end plate failure

$$\phi_p \cdot M_{cp} = \phi_p \cdot \left\{ \frac{t_p^2 \cdot f_p \cdot \left(w_{eq} \cdot (d' + 2 \cdot s_o') + (w_{eq} - n \cdot d_f) \cdot d'\right) + n \cdot \dfrac{\pi \cdot d_b^3 \cdot f_{yb}}{8} \cdot d'}{4 \cdot d' \cdot s_o'} \right\} \cdot (d - t_s) \quad (9.7)$$

Alternatively, for a given connection design moment $M^*$, the appropriate end plate thickness ($t_{pu}$) is given by

$$t_{pu} = 2 \cdot \sqrt{\frac{\left\{ \dfrac{M^* \cdot s_o'}{\phi_p \cdot (d + t_s)} - n \cdot \dfrac{\pi \cdot d_b^3 \cdot f_{yb}}{32} \right\} \cdot d'}{\left\{ w_{eq} \cdot (d' + 2 \cdot s_o') + (w_{eq} - n \cdot d_f) \cdot d' \right\} \cdot f_p}} \quad (9.8)$$

For serviceability limit state design based on bolt yielding

$$\phi_b \cdot M_{cbs} = \phi_b \cdot \left\{ \frac{\left(d' + 2 \cdot (s_o' + a_e)\right) \cdot w_{eq} \cdot t_p^2 \cdot f_y}{4 \cdot (s_o' + a_e) \cdot d'} + \frac{n \cdot B_{yl} \cdot a_e}{s_o' + a_e} \right\} \cdot (d - t_s) \quad (9.9)$$

Alternatively, if the connection design serviceability moment $M_s^*$ is known, the appropriate end-plate thickness is given by

$$t_{bs} = 2 \cdot \sqrt{\frac{\dfrac{M_s^* \cdot (s_o' + a_e) \cdot d'}{\phi_b \cdot (d - t_s)} - n \cdot B_{yl} \cdot a_e \cdot d'}{\left(d' + 2 \cdot (s_o' + a_e)\right) \cdot w_{eq} \cdot f_y}} \quad (9.10)$$

For serviceability limit state design based on plate yielding

$$\phi_p \cdot M_{cps} = \phi_p \cdot \left\{ \frac{\left((d' + s_o') \cdot w_{eq} - n \cdot d_f \cdot d'\right) \cdot t_p^2 \cdot f_y}{2 \cdot s_o' \cdot d'} \right\} \cdot (d - t_s) \quad (9.11)$$

Alternatively, if the connection design serviceability moment $M_s^*$ is known, the appropriate end-plate thickness is given by

$$t_{ps} = \sqrt{\frac{2 \cdot M_s^* \cdot s_o' \cdot d'}{\phi_p \cdot (d - t_s) \cdot \left((d' + s_o') \cdot w_{eq} - n \cdot d_f \cdot d'\right) \cdot f_y}} \tag{9.12}$$

The dimensions ($a_e$, $a_p$, $d$, $d'$, $s_o'$, $t_p$, $t_s$, $d_f$) are defined in Figure 9.5. The capacity factors are given in AS 4100-1998 as $\phi_b = 0.8$ and $\phi_p = 0.9$. $n$ is the number of bolts, $d_b$ is the bolt diameter, $w_{eq}$ is the equivalent plate width which is the same as the end plate width $W_p$ if bolts are placed in line with webs, $B_{ul}$ is the tensile strength of individual bolt, $B_{yl}$ is the yield load of individual bolt, $f_y$ is the plate yield stress, $f_{yb}$ is the bolt yield stress, $f_p$ is the plate design stress defined as $f_p = (f_y+2f_u)/3$ where $f_u$ is the plate tensile strength.

### 9.3.4 *Design Procedures*

Design procedures proposed by Wheeler et al (1998) are summarised as follows:

Step 1 Estimate the end plate dimensions for initial design based on section size, bolt size, and number of bolts.

Step 2 If two or more bolts are not positioned within the webs of the section ($c \leq 0$), yield line analysis is required to determine the equivalent width ($w_{eq}$). Otherwise, the equivalent width is equal to the end plate width ($W_p$). If the plate thickness is already known, go to step 6.

Step 3 Solve for the strength limit state design thicknesses $t_{bu}$ and $t_{pu}$ using Equation (9.5) and Equation (9.8). For appropriate ultimate strength limit state design, the required plate thickness ($t_u$) is equal to the maximum of the thickness based on bolt and plate capacity.

$$t_u = \max(t_{bu}, t_{pu})$$

Step 4 Solve for the serviceability design thicknesses $t_{bs}$ and $t_{ps}$ using Equation (9.10) and Equation (9.12). For appropriate serviceability limit state design, the required plate thickness ($t_{sv}$) is equal to the maximum of the serviceability thicknesses calculated.

$$t_{sv} = \max(t_{bs}, t_{ps})$$

Step 5 The resulting thickness for the end plate ($t_p$) must exceed both the serviceability and ultimate limit state thicknesses, but must be less than the maximum allowable plate thickness ($t_{max}$) given by Equation (9.6). That is

$$\max(t_{sv}, t_u) \leq t_p \leq t_{max}$$

Step 6 Solve Equation (9.4) and Equation (9.7) using $t_p$ to obtain the design moment capacities of the connection ($\phi_b M_{cb}$ *and* $\phi_p M_{cp}$). If $\phi_b M_{cb} > \phi_p M_{cp}$, select either an alternative bolting arrangement to lower the bolt capacity or a thicker plate to increase the plate capacity. Similarly, if $\phi_b M_{cb} < M^*$, the bolt capacity must be increased. Moment capacities should then be recalculated to ensure that they exceed the design moment.

Step 7 Solve Equation (9.9) and Equation (9.11) to ensure that the serviceability limit moments $\phi_b M_{cbs}$ and $\phi_p M_{cps}$ are greater than the serviceability design moment $M_s^*$.

Step 8 The serviceability limit moment is the minimum moment from step 7, and the connection design capacity for the strength limit state is the minimum moment from step 6.

Three design examples can be found in Wheeler et al (1997).

## 9.4 Plastic Design of Portal Frames

### *9.4.1 General*

Plastic design of statically indeterminate frames can lead to higher ultimate loads with associated higher deformations compared to traditional elastic design methods. As some point in the frame reaches its plastic moment ($M_p$), a plastic zone is created. The zone forms a hinge and rotates further, maintaining $M_p$, while redistributing additional load to other parts of the structure. The hinge process is repeated as other hinges form, until there is a sufficient number of hinges to create a plastic collapse mechanism for the whole structure or part of the structure. All hinges, particularly those which form early, must be able to rotate sufficiently for this mechanism to form.

There are specific requirements for the suitability of sections for plastic design. A Compact or Class 1 section is deemed suitable for plastic design. The flanges and webs of the RHS must be sufficiently stocky to avoid local buckling before large plastic rotations occur. Appropriate slenderness limits for cold-formed RHS were given in Chapter 3. The rotational capacity of knee joints was examined by Wilkinson and Hancock (2000), and Teh and Hancock (2004) since plastic hinges often form at the connections of portal frames.

Due to the large strains occurring in plastic hinges, material ductility requirements are imposed by design standards. Most cold-formed RHS do not satisfy the requirements of Clause 5.2.3.3 of BS 5950 Part 1 or Clause 4.5.2 of AS 4100. This section briefly describes tests on portal frames manufactured from cold-formed RHS and the improved knee joints followed by some conclusions on the suitability of cold-formed RHS for plastic design.

### *9.4.2 Portal Frame Tests*

Three frames were tested under simulated combined gravity and transverse wind loads (Wilkinson and Hancock 2000). The sizes of RHS and the ratio of vertical load ($V$) to horizontal load ($H$) are summarised in Table 9.9. The general layout of each frame is shown in Figure 9.7. Each frame spanned 7 metres, with an eaves height of 3 metres, and a total height of 4 metres. There was a collar tie for loading which joined the midpoint of each rafter. The frames were constructed from RHS, and the collar tie was made from a pair of channel sections.

Table 9.9 Results of portal frame tests (Wilkinson and Hancock 2000)

| Specimen | RHS used | Measured yield stress of RHS ($N/mm^2$) | Ratio of vertical to horizontal load ($V/H$) | Ultimate Load (kN) | |
|---|---|---|---|---|---|
| | | | | Vertical | Horizontal |
| Frame 1 | 150 × 50 × 4.0 RHS C350 | 411 | 40 | 68.2 | 1.75 |
| Frame 2 | 150 × 50 × 4.0 RHS C450 | 438 | 40 | 71.5 | 1.87 |
| Frame 3 | 150 × 50 × 4.0 RHS C450 | 438 | 3.3 | 45.7 | 13.8 |

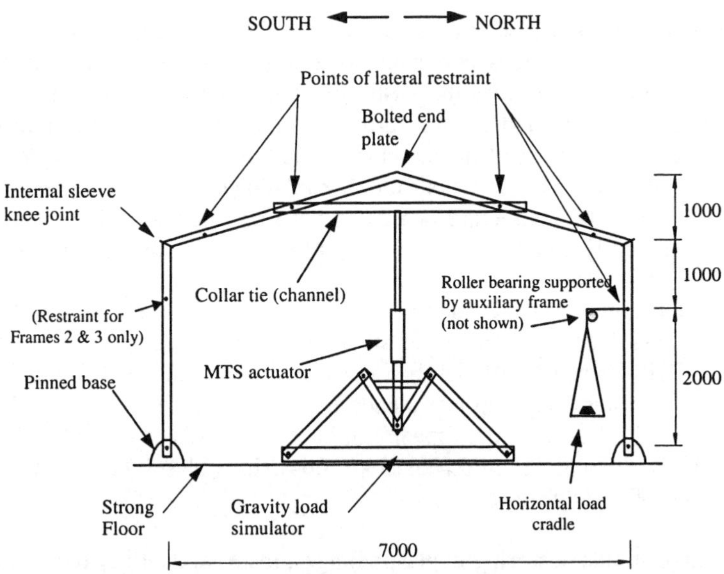


Figure 9.7 Portal frame test arrangement (Wilkinson and Hancock 2000)

A combination of vertical gravity and transverse wind loads was selected for the test series. In a prototype portal frame, gravity and wind loads are transferred to the frame from the sheeting via purlins and girts, similar to a distributed load. It is difficult to replicate distributed loads in a laboratory, and hence the loads were applied as point loads.

The vertical load was applied at the midpoint of the collar tie. Large in-plane sway displacements were anticipated and it was essential that the downwards loads remained vertical to simulate true gravity loading. A hydraulic jack was required to achieve the large vertical load. However, a jack connected directly to the strong floor of the laboratory would produce a non-vertical load as the frame swayed horizontally. A gravity load simulator was attached to the strong floor of the laboratory as shown in

Figure 9.7. The simulator had two inclined members, pinned together by a rigid triangular unit, and pinned at the other end to the base. The jack was bolted to the top of another vertical member, which was pinned to the base of the triangular section.

The welded internal sleeve joint (Wilkinson and Hancock 2000), as shown in Figure 9.8 was selected for the knee joints of the portal frame. A welded box section, cut in an open "V" shape (to match the angle of the portal frame knee), was fabricated from 10 mm Grade 250 plate. The sleeve was sized to fit tightly inside the RHS. The sleeve extended 300 mm into both the column and the rafter of each knee joint and required a sledge hammer to insert the sleeve in place. The sleeve connection forces the plastic hinge at the knee to form away from the connection centreline, as shown in Figure 9.9.

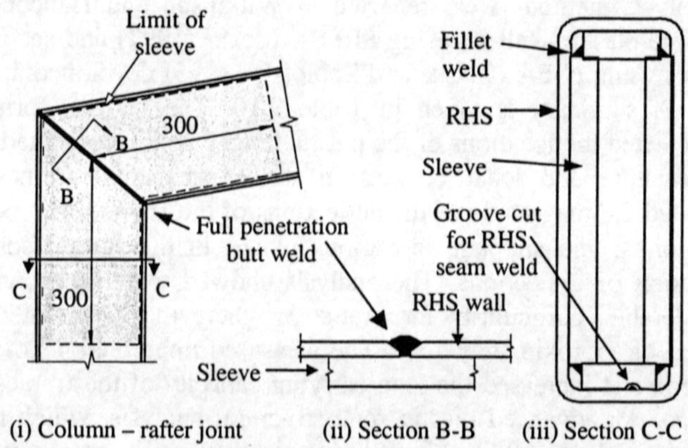


Figure 9.8 Internal sleeve knee joint connection

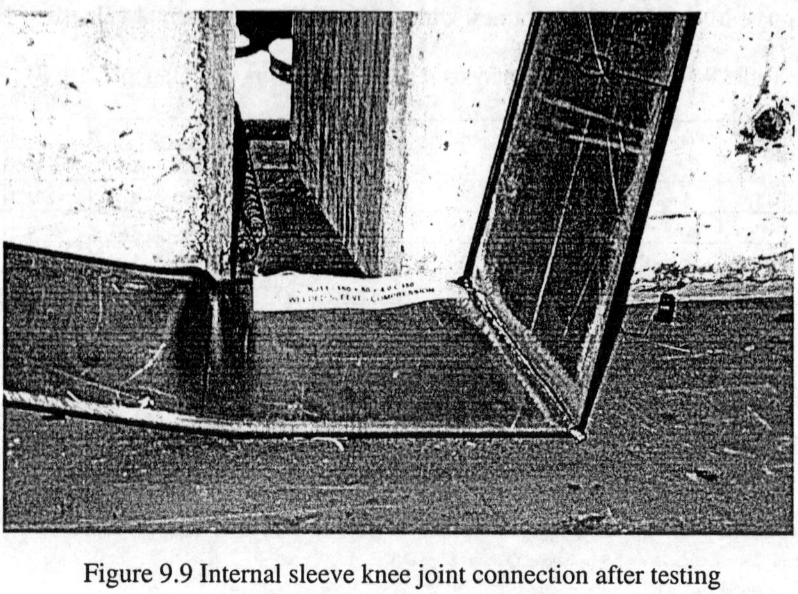

Figure 9.9 Internal sleeve knee joint connection after testing (Wilkinson and Hancock 2000)

A moment resisting bolted end plate connection was used at the apex of the frame. A 10 mm plate was butt welded to the end of each rafter, and the two plates were bolted together with eight high strength fully-tensioned M16 bolts positioned symmetrically about the $x$-axis of the RHS.

A common connection for portal frames is the bolted base plate connection to a concrete footing. To simulate the column base plate, a pinned connection was used. The webs of the RHS were stiffened locally by a steel plate, and a 30 mm diameter high strength steel pin was inserted through the neutral axis of the RHS. A thick steel base plate supported the pin. The base plate was securely fastened to the strong floor of the laboratory. The pin connection was greased to reduce friction.

Various analysis methods were reported in Wilkinson and Hancock (1999). They include simple plastic analysis using PRFSA (CASE 1997) and second order plastic zone analysis using NIFA (Clarke and Zablotskii 1995) developed at The University of Sydney. A summary is given in Table 9.10. The various forms of structural analyses predicted the locations of the plastic hinges which coincided with the places of high curvature and local buckles in the experimental frames. All analyses underestimated the magnitude of the deflections of the frame. Factors such as loss of rigidity in joints, and slippage in various places in the frame, contributed to the underestimation of deflections. The analysis showed that the second order effects were considerable, particularly for Frame 3, where they accounted for a drop in ultimate load of approximately 15%. The measured imperfections of the frame were unsympathetic and increased the load carrying capacity of the frames compared to a perfect frame. An advanced plastic zone structural analysis, which included second order effects, gradual yielding, multilinear stress-strain curves varying around the cross-section, and the structural imperfections slightly overestimated the strength of the frame. The main cause for the over prediction of strength was the underestimation of the deflections and hence the magnitude of the second order effects. The second order plastic zone analysis without frame imperfections provided the best estimates of the ultimate loads of the three frames, within 2% of the experimental values.

Table 9.10 Summary of analysis results (Wilkinson and Hancock 1999)

| Program | Type[1] | Properties[2] | Materials[3] | Imperfection[4] | Frame 1 | | Frame 2 | | Frame 3 | |
|---|---|---|---|---|---|---|---|---|---|---|
| | | | | | Vertical[5] | Ratio[6] | Vertical[5] | Ratio[6] | Vertical[5] | Ratio[6] |
| PRFSA | 1st pl | nominal | e-p | N | 57.64 | 0.85 | 74.03 | 1.04 | 50.82 | 1.11 |
| PRFSA | 1st pl | measured | e-p | N | 68.41 | 1.00 | 72.76 | 1.02 | 49.96 | 1.09 |
| NIFA | 1st pz | nominal | e-p | N | 57.18 | 0.84 | 73.51 | 1.03 | 48.94 | 1.07 |
| NIFA | 1st pz | measured | e-p | N | 66.89 | 0.98 | 70.85 | 0.99 | 47.12 | 1.03 |
| NIFA | 2nd pz | measured | e-p | N | 61.49 | 0.90 | 64.79 | 0.91 | 40.12 | 0.88 |
| NIFA | 2nd pz | measured | mul | N | 67.16 | 0.98 | 73.14 | 1.02 | 45.71 | 1.00 |
| NIFA | 2nd pz | measured | mul | Y | 68.84 | 1.01 | 77.05 | 1.08 | 46.63 | 1.02 |

Note:

(1) Type of analysis is either 1st order simple plastic (1st pl), 1st order plastic zone (1st pz), or 2nd order plastic zone (2nd pz).

(2) Dimensions and yield stress are based on with nominal or measured properties.

(3) Material properties are wither elastic-plastic (e-p) with the properties of the webs used for the entire cross section, or multi-linear (mul) approximation to the measured properties with separate properties for the webs, flanges and corners.

(4) The frame imperfections are ether included (Y) or excluded (N).

(5) Ultimate vertical load in kN.

(6) Ratio of the analysis to the experimental load.

### 9.4.3 *Improved Knee Joints*

A stiffened knee joint between cold-formed DuraGal C450 RHS welded in the normal way does not have a sufficient rotation capacity for plastic design under opening moment, unless internal sleeves or similar other reinforcement means are employed to prevent premature fracture of the connection in the corner adjacent to the stiffening plate (Wilkinson 1999). Under opening moment there are highly pronounced strain concentrations in the tension corners at the connected end of the RHS. Coupled with the low ductility of the highly cold-worked corners and the reduced ductility of the heat affected zone, a normal stiffened welded knee joint between DuraGal RHS fractures prematurely in one of the tension corners.

A simple and effective solution has been proposed by Teh and Hancock (2004). The solution involves depositing extra layers of weld on the tension flanges (see Figure 9.10) so that strain concentrations are shifted away from the brittle corners and the greatest strain at a given rotation is considerably lower than that would otherwise be incurred. As a result, fracture of a stiffened welded knee joint having such extra layers of weld takes place in the more ductile flange rather than in the corner, and at a greater rotation of the plastic hinge. The extra layers of weld can be deposited using the welding machine used to produce the connection weld, without changing the machine settings and the welding consumables.

It was found through laboratory tests that the solution increased the rotation capacities of stiffened welded knee joints between DuraGal C450 150 × 50 × 4 and 150 × 50 × 5 RHS to the extent that the minimum rotation capacity required for plastic design was surpassed. The proposed solution is not only more economical than the use of internal sleeves but is also applicable to existing structures provided the joints are accessible to depositing the extra layers of weld. This feature of the proposed solution means that the load-carrying capacity of an existing portal frame may be upgraded provided it is predicted to fail by a plastic collapse mechanism.

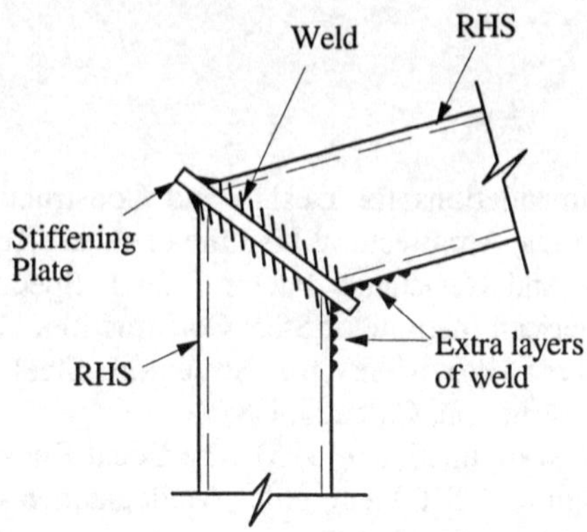


Figure 9.10 Improve knee joint (Teh and Hancock 2004)

### 9.4.4 *Summary*

Detailed work on plastic behaviour of cold-formed rectangular hollow sections can be found in Wilkinson (1999) where the following conclusions were made:

1. While cold-formed RHS do not satisfy the material ductility requirements specified for plastic design in some current steel design standards, plastic hinges and plastic collapse mechanisms formed. This suggests that the restriction on plastic design for cold-formed RHS based on insufficient material ductility is unnecessary, provided that the connections are suitable for plastic hinge formation, if required.
2. Cold-formed RHS can be used in plastic design, but stricter element slenderness limits and consideration of the connections, are required.

## 9.5 Other Recent Developments

Other recent developments in cold-formed tubular members and connections are listed below with some references given.

- Fracture toughness of RHS (Kosteski et al 2005)
- Very high strength steel tubes in tension, compression and bending (Zhao 2000, Jiao and Zhao 2001, 2003, 2004a, 2004b), and the utilisation of the high yield stress of steel tubes through fabricated sections (Zhao et al 2004, Binh et al 2004, Rhodes et al 2005) or CFRP (Carbon Fibre Reinforced Polymer) strengthening (Jiao and Zhao 2004c).
- Bird beak T-joints (Owen et al 1996, Davies et al 2001, Keizer et al 2003)
- Welding in corners of cold-formed RHS (Puthli et al 2004) and butt-welded connections between equal-width RHS (Teh and Rasmussen 2002)
- Bolted end plate connections joining rectangular hollow sections using 8 bolts (Wheeler et al 2003)
- Gusset plate connections (Korol 1996, Cheng et al 1998, Cheng and Kulak 2000, Willibald et al 2003, 2004, Wilkinson et al 2002, Ling et al 2004)
- Concrete Filled Double Skin Tubes (CFDST) (Zhao and Grzebieta 2002, Zhao et al 2002a, 2002b, Elchalakani et al 2002, Han et al 2004, Tao et al 2004).
- Fire resistance (Ala-Outinen and Myllymäki 1995, Patterson et al 1999, Zhao et al 2001, Han et al 2003)

## 9.6 References


1. AIJ (1997), Recommendations for Design and Construction of Concrete-Filled Steel Tubular Structures, Architectural Institute of Japan, Tokyo, Japan
2. AISC (1999), Load and Resistance Factor Design Specification for Structural Steel Buildings, American Institute of Steel Construction, Chicago, Illinois, USA
3. AISC (2002), Seismic Provisions for Structural Steel Buildings, American Institute of Steel Construction, Chicago, USA
4. Ala-Outinen, T. and Myllymäki, J. (1995), The Local Buckling of RHS Members at Elevated Temperatures, VTT Tiedotteita - Meddelanden - *Research Notes 1672*, VTT Building Technology, Finland
5. Berge, S. and Webster, S.E. (1987), The Size Effect on the Fatigue Behaviour of Welded Joints, *Proceedings*, Steel in Marine Structures, (SIM'87), pp 179-203
6. Bergmann, R., Matsui, C., Meinsma, C. and Dutta, D. (1995), *Design Guide for Concrete Filled Hollow Section Columns under Static and Seismic Loading*, TÜV-Verlag, Köln, Germany

7. Binh, D.V., Al-Mahaidi, R. and Zhao, X.L. (2004), Finite Element Analysis of Fabricated and Triangular Section Stub Columns Utilizing Very High Strength Steel Tubes, *Advances in Structural Engineering – An International Journal,* 7(5), pp 447-460
8. Birkemoe, P.C. and Packer, J.A. (1986), Ultimate Strength Design of Bolted Tubular Tension Connections, *Proceedings*, Steel Structures – Recent Research Advances and their Applications to Design, Budva, Yugoslavia, pp 153-168
9. BSI (2000), Structural Use of Steelwork in Building, BS 5950, Part 1, British Standard Institution, London, UK
10. Cao, J.J., Packer, J.A and Du, S. (2000), Finite Element Modeling of Bolted Flange Connections, In: *Structural Failure and Plasticity*, Zhao, X.L. and Grzebieta, R.H. (eds), Elsevier: Oxford, pp 473-478
11. CASE (1997), PRFSA: Plane Rigid Frame Structural Analysis, *User Manual*, Centre for Advanced Structural Engineering, The University of Sydney, Sydney, Australia
12. Cheng, J.J.R., Kulak, G.L. and Khoo, H. (1998), Strength of Slotted Tubular Tension Members, *Canadian Journal of Civil Engineering*, 25(6), pp 982-991
13. Cheng, J.J.R. & Kulak, G.L. (2000), Gusset Plate Connection to Round HSS Tension Members. *Engineering Journal,* American Institute of Steel Construction, $4^{th}$ Quarter, pp 133-139
14. Clarke, M.J. and Zablotskii, S.V. (1995), NIFA: Non-linear Inelastic Frame Analysis, *User Manual,* Centre for Advanced Structural Engineering, The University of Sydney, Sydney, Australia
15. Davies, G., Owen, J.S. and Kelly, R. (2001), The Effect of Purlin Loads on the Capacity of Overlapped Bird-Beak K-joints, In: *Tubular Structures IX,* Puthli, R.S. and Herion, S. (eds), Swets & Zeitlinger: Lisse, pp 229-238
16. DEn (1993), Background to New Fatigue Design Guidance for Steel Joints in Offshore Structures, *Internal Report*, Department of Energy, London, UK
17. Dijkstra, O.D., van Foeken, R.J., Romeijn, A., Karamanos, S.A., van Wingerde, A.M., Puthli, R.S., Herion, H. and Wardenier, J. (1996), Fatigue Design Guide for Circular and Rectangular Hollow Section Multiplanar Joints, *TNO-Report*, 97-CON-R1331, Delft, The Netherlands
18. Dimitrakis, S.D., Lawrence, F.V. and Mohr, W.C. (1995), S-N Curves for Welded Tubular Joints, *Proceedings*, Fourteenth International Offshore Mechanics and Arctic Engineering Symposium (OMAE'95), Volume III, pp 209-222
19. EC3 (2003), Eurocode 3: Design of Steel Structures – Part 1.1: General Rules and Rules for Buildings, prEN 1993-1-1:2003, November 2003, European Committee for Standardization, Brussels, Belgium
20. EC 4 (2001), Eurocode 4 Part 1.1, Design of Composite Steel and Concrete Structures: General rules and rules for buildings, prEN 1994-1-1:2001, April 2001, European Committee for Standardization, Brussels, Belgium
21. EC8 (2003), Eorocode 8 Part 1.1, Design of Structures for Earthquake Resistance: General Rules, seismic actions and rules for buildings, European Committee for Standardization, Brussels, Belgium
22. Efthymiou, M. (1988), Development of SCF Formulae and Generalized Functions for use in Fatigue Analysis, *Proceedings*, Offshore Tubular Joints Conference (OTJ'88) on Recent Developments in Tubular Joints Technology, Surrey, UK, 4-5 October

23. Efthymiou, M. and Durkin, S. (1985), Stress Concentration in T/Y and Gap/Overlap K-joints, In: *Behaviour of Offshore Structures*, Elsevier Science Publishers, Amsterdam, The Netherlands, pp 429-440
24. Elchalakani, M., Zhao, X.L. and Grzebieta, R.H. (2001), Concrete Filled Circular Steel Tubes subjected to Pure Bending, *Journal of Constructional Steel Research*, 57(11), pp 1141-1168
25. Elchalakani, M., Zhao, X.L., and Grzebieta, R.H. (2002), Tests on Concrete Filled Double Skin (CHS outer and SHS inner) Composite Short Columns under Axial Compression, *Thin-Walled Structures*, 40(5), pp 415-441
26. Elchalakani, M. (2003), Cyclic Bending Behaviour of Hollow and Concrete-Filled Cold-formed Circular Steel Members, *PhD Thesis*, Monash University, Melbourne, Australia
27. Elchalakani, M., Zhao, X.L., and Grzebieta, R.H. (2003), Tests of Cold-Formed Circular Tubular Braces under Cyclic Axial Loading, *Journal of Structural Engineering,* ASCE, 129(4), pp 507-514
28. Elchalakani, M., Zhao, X.L., and Grzebieta, R.H. (2004a), Cyclic Bending Tests to Determine Fully Ductile Section Slenderness Limits for Cold-Formed CHS, *Journal of Structural Engineering,* ASCE, 130(7), pp 1001-1010
29. Elchalakani, M., Zhao, X.L., and Grzebieta, R.H. (2004b), Concrete-Filled Steel Tubes Subjected to Constant Amplitude Cyclic Pure Bending, *Engineering Structures*, 26(14), pp 2125-2135
30. Gurney, T.R. (1979), *Fatigue of Welded Structures*, Cambridge University Press, 2nd Edition, Cambridge, UK
31. Haagensen, P.J. (1989), Improvement Techniques, *Proceedings*, International Symposium on the Occasion of the Retirement of Prof. J de Back, Delft, The Netherlands, pp 77-89
32. Han, L.H, Yao, G.H. and Zhao, X.L. (2004), Concrete-Filled Double Skin (SHS outer and CHS inner) Steel Tubular Beam-Columns, *Thin-Walled Structures*, 42(9), pp 1329-1355
33. Han L.H., Zhao, X.L, Yang, Y.F. and Feng, J.B. (2003), Experimental Study and Calculation of Fire Resistance of Concrete-Filled Hollow Section Columns, *Journal of Structural Engineering,* ASCE, 129(3), pp 345-356
34. Igarashi, S., Wakiyama, K., Inoue, K., Matsumoto, T., Murase, Y. (1985), Limit Design of High Strength Bolted Tube Flange Joint, *Journal of Structural and Construction Engineering*, Transactions of AIJ, Osaka University, Japan
35. IIW (1985), Recommended Fatigue Design Procedure for Hollow Section Joints. Part 1 – hot spot stress method for nodal joints. *IIW Doc. XV-582-85* and *IIW Doc. XIII-1158-85*, IIW Assembly, Strasbourg, France
36. IIW (1999), *Fatigue Design Procedure for Welded Hollow Section Joints*, Part 1: Recommendations and Part 2: Commentary, Zhao, X.L. and Packer, J.A. (eds), Abington Publishing, Cambridge, UK
37. Jiao, H. and Zhao, X.L. (2001), Material Ductility of Very High Strength (VHS) Circular Steel Tubes in Tension, *Thin-Walled Structures*, 39 (11), pp 887-906
38. Jiao, H. and Zhao, X.L. (2003), Imperfection, Residual Stress and Yield Slenderness Limit of Very High Strength (VHS) Circular Steel Tubes, *Journal of Constructional Steel Research*, 59(2), pp 233-249
39. Jiao, H. and Zhao, X.L. (2004a), Tension Capacity of Very High Strength (VHS) Circular Steel Tubes after Welding, *Advances in Structural Engineering – An International Journal,* 7(4), pp 85-96

40. Jiao, H. and Zhao, X.L. (2004b), Section Slenderness Limits of Very High Strength Circular Steel Tubes in Bending, *Thin-Walled Structures,* 42(9), pp 1257-1271
41. Jiao, H. and Zhao, X.L. (2004c), CFRP Strengthened Butt-Welded Very High Strength (VHS) Circular Steel Tubes, *Thin-Walled Structures,* 42(7), pp 963-978
42. Karamanos, S.A., Romeijn, A. and Wardenier, J. (1997), Stress Concentrations and Joint Flexibility Effects in Multiplanar Welded Tubular Connections for Fatigue Design, Stevin Report 6-98-05, *CIDECT Report 7R-17/98*, Delft University of Technology, Delft, The Netherlands
43. Kato, B. and Hirose, A. (1984), Bolted Tension Flanges Joining Circular Hollow Section Members, *CIDECT Report 8C-84-24-E*, University of Tokyo, Tokyo, Japan
44. Kato, B. and Mukai, A. (1982), Bolted Tension Flanges Joining Square Hollow Section Members, *CIDECT Report 8B-82/3-E*, University of Tokyo, Tokyo, Japan
45. Keizer, R., Romeijn, A., Wardenier, J. and Glijnis, P.C. (2003), The Fatigue Behaviour of Diamond Bird Beak T-joints, In: *Tubular Structures X*, Jaurrieta, M.A., Alonso, A. and Chica, J.A. (eds), Balkema: Rotterdam, pp 303-309
46. Korol, R.M. (1996), Shear Lag in Slotted HSS Tension Members. *Canadian Journal of Civil Engineering,* Vol. 23, pp 1350-1354
47. Kosteski, N., Packer, J.A. and Puthli, R.S. (2005), Notch Toughness of Internationally Produced Hollow Sections, *Journal of Structural Engineering,* ASCE, 131(2), pp 279-286
48. Ling, T.W., Zhao, X.L., Al-Mahaidi, R. and Packer, J.A. (2004), Connection Design of Very High Strength Steel Tubes Longitudinally Welded to Steel Plates, In: *Developments in Mechanics of Structures and Materials*, Deeks, A.J. and Hao, H. (eds), Balkema: London, pp 1135-1140
49. Mang, F. (1980), Investigation of Standard Bolted Flange Connections for Circular and Rectangular Hollow Sections, *CIDECT Report 8A-81/7-E*, University of Karlsruhe, Karlsruhe, Germany
50. Marshall, P.W. (1984), Connections for Welded Tubular Structures, *Proceedings*, 2nd International Conference on Welding of Tubular Structures, Boston, USA, pp 1-54
51. Marshall, P.W. (1992), *Design of Welded Tubular Connections – Basis and Use of AWS Code Provisions*, Elsevier Science Publishers, Amsterdam, The Netherlands
52. Mashiri, F.R., Zhao, X.L. and Grundy, P. (2001), Effects of Weld Profile and Undercut on Fatigue Crack Propagation Life of Thin-Walled Cruciform Joint, *Thin-Walled Structures*, 39(3), pp 261-285
53. Mashiri, F.R., Zhao, X.L. and Grundy, P. (2002a), Fatigue Tests and Design of Thin Cold-Formed SHS-to-SHS T-Connections under In-Plane Bending, *Journal of Structural Engineering*, ASCE, 128 (1), pp 22-31
54. Mashiri, F.R., Zhao, X.L. and Grundy, P. (2002b), Fatigue Tests and Design of Welded T Connections in Thin Cold-Formed Square Hollow Sections under In-Plane Bending, *Journal of Structural Engineering*, ASCE, 128 (11), pp 1413-1422
55. Mashiri, F.R., Zhao, X.L. and Grundy, P. (2004), Stress Concentration Factors and Fatigue Failure of Welded T-connections in Thin Cold-Formed Circular Hollow Sections under In-plane Bending, *International Journal of Structural Stability and Dynamics*, 4(3), pp 403-422

56. Matsui, C., Mitani, I., Kawano, A. and Tsuda, K. (1997), AIJ Design Method for Concrete Filled Steel Tubular Structures, *Proceedings*, ASCCS Seminar on Concrete Filled Steel Tubes – A Comparison of International Codes and Practice, September, Innsbruck, Austria, pp 93-116
57. NZS 3404 (1992), Steel Structures Standard. Standards Association of New Zealand
58. Owen, J.S., Davies, G. and Kelly, R. (1996), A Comparison of the Behaviour of RHS Bird Beak T-joints with Normal RHS and CHS Systems, In: *Tubular Structures VII*, Farkas, J. and Jármai, H. (eds), Balkema: Rotterdam, The Netherlands, pp 173-180
59. Packer, J.A., Bruno, L. and Birkemoe, P.C. (1989), Limit Analysis of Bolted RHS Flange Plate Joints, *Journal of Structural Engineering,* ASCE, 115(9), pp 2226-2242
60. Packer, J.A., Wardenier, J., Kurobane, Y., Dutta, D. and Yeomans, N. (1992), *Design Guide for Rectangular Hollow Section (RHS) Joints under Predominantly Static Loading*, TÜV-Verlag, Köln, Germany
61. Packer, J.A. and Henderson, J.E. (1997), *Hollow Structural Section Connections and Trusses*, Canadian Institute of Steel Construction, Ontario, Canada
62. Packer, J.A., Du. S. and Willibald, S. (1999), Bolted Flange-Plate Connections for RHS Tension Members, *CIDECT Report 8D-12/99*, April, University of Toronto, Toronto, Canada
63. Patterson, N.L., Zhao, X.L., Wong, B., Ghojel, J. and Grundy, P. (1999), Elevated Temperature Testing of Composite Columns, In: *Advances in Steel Structures*, Chan, S.L. and Teng, J.G. (eds), Elsevier: Oxford, pp 1047-1054
64. Puthli, R.S., de Koning, C.H.M., van Wingerde, A.M., Wardenier, J. and Dutta, D. (1989), Fatigue Strength of Welded Unstiffened RHS Joints in Lattice Structures and Vierendeel Girders, Final Report Part III: Evaluation for Design Rules, *TNO-IBBC Report No. BI-89-097/63.5.3800*, Stevin Report No. 25-6-89-36/A1, June, Delft, The Netherlands
65. Puthli, R.S. and Herion, S. (1996), Stress Concentration and Secondary Moment Distribution in RHS Joints for Fatigue Design, *Background Document* for fatigue design guide on RHS, University of Karlsruhe, Karlsruhe, Germany
66. Puthli, R.S., Herion, S., Boellinghaus, T. and Florian, N.W. (2004), Welding in Cold-Formed Areas of Rectangular Hollow Sections, *CIDECT Report 1A-5/04*, University of Karlsruhe, Karlsruhe, Germany
67. Ranzi, G. and Kneen, P. (2002), Design of Pinned Columns Base Plates, *Steel Construction*, Journal of the Australian Steel Institute, 36(2), pp 1-53
68. Rhodes, J., Zhao, X.L., Binh, D.V. and Al-Mahaidi, R. (2005), Rational Design Analysis of Stub Columns Fabricated using Very High Strength Circular Steel Tubes, *Thin-Walled Structures*, 43(3), pp 445-460
69. Romeijn, A., Wardenier, J., de Koning, C.H.M., Puthli, R.S. and Dutta, D. (1993), Fatigue Behaviour and Influence of Repair on Multiplanar K-joints Made of Circular Hollow Sections, *Proceedings*, 3rd International Offshore and Polar Engineering Conference, Singapore, Vol. IV, pp 27-36
70. Romeijn, A. (1994), Stress and Strain Concentration Factors of Welded Multiplanar Tubular Joints, *PhD Thesis*, Delft University of Technology, Delft, The Netherlands
71. Standards Australia (1998), Steel Structures, Australian Standard AS 4100, Standards Australia, Sydney, Australia

72. Tao, Z., Han, L.H. and Zhao, X.L. (2004), Behaviour of Concrete-Filled Double Skin (CHS inner and CHS outer) Steel Tubular Stub Columns and Beam-Columns, *Journal of Constructional Steel Research*, 60(8), pp 1109-1256
73. Teh, L.H. and Rasmussen, K.J.R. (2002), Strength of Butt Welded Connections between Equal-Width Rectangular Hollow Sections, *Research Report R817*, The University of Sydney, Sydney, Australia
74. Teh, L.H. and Hancock, G.J. (2004), Improving the Ductility of Knee Joints between Rectangular Hollow Sections, *Journal of Structural Engineering*, ASCE, 130(11), pp 1790-1798
75. Thorpe, T.W. and Sharp, J.V. (1989), The Fatigue Performance of Tubular Joints in Air and Sea Water, *MaTSU Report*, Harwell Laboratory, Oxfordshire, UK
76. Uy, B. (2000), Strength of Concrete Filled Steel Box Columns Incorporating Local Buckling, *Journal of Structural Engineering,* ASCE, 126(3), pp 341-352
77. van Delft, D.R.V. (1981), A Two Dimensional Analysis of the Stresses at the Vicinity of the Weld Toes of Tubular Structures, *Stevin Report 6-18-8*, Delft University of Technology, Delft, The Netherlands
78. van Delft, D.R.V., Noordhoek, C. and de Back, J. (1985), Evaluation of the European Fatigue Test Data on Large-Sized Welded Tubular Joints for Offshore Structures, *Proceedings*, Offshore Technology Conference, Houston, USA, paper OTC 4999
79. van Wingerde, A.M. (1992), The Fatigue Behaviour of T and X Joints Made of Square Hollow Sections, *Heron*, The Netherlands, 37(2), pp 1-180
80. van Wingerde, A.M., Packer, J.A., Wardenier, J. and Dutta, D. (1996), The Fatigue Behaviour of K-joints Made of Square Hollow Sections, *CIDECT Report 7P-19/96*, University of Toronto, Toronto, Canada
81. van Wingerde, A.M., Packer, J.A., Wardenier, J. and Dutta, D. (1997a), Simplified Design Graphs for the Fatigue Design of Multiplanar K-joints with Gap, *CIDECT Report 7R-01/97*, January, University of Toronto, Toronto, Canada
82. van Wingerde, A.M., Packer, J.A. and Wardenier, J. (1997b), IIW Fatigue Rules for Tubular Joints, *Proceedings*, IIW International Conference on Performance of Dynamically Loaded Welded Structures, July, San Francisco, USA, pp 98-107
83. van Wingerde, A.M., van Delft, D.R.V., Wardenier, J. and Packer, J.A. (1997c), Scale Effects on the Fatigue Behaviour of Tubular Structures, *Proceedings*, IIW International Conference on Performance of Dynamically Loaded Welded Structures, July, San Francisco, USA, pp 123-135
84. van Wingerde, A.M., Wardenier, J. and Packer, J. A. (1998a), Commentary on the Draft Specification for Fatigue Design of Hollow Section Joints, In: *Tubular Structures XIII*, Choo, S. and van der Vegte, G.J. (eds), Balkema, Rotterdam, The Netherlands, pp 117-127
85. van Wingerde, A. M., Wardenier, J. and Packer, J.A. (1998b), Simplified Design Graphs for the Fatigue Design of Multiplanar K-joints with Gap, *CIDECT Final Report No. 7R-23/98*, September, Delft University of Technology, Delft, The Netherlands
86. Wardenier, J. (1982), *Hollow Section Joints*, Delft University Press, Delft, The Netherlands
87. Wardenier, J., Kurobane, Y., Packer, J.A., Dutta, D. and Yeomans, N. (1991), *Design Guide for Circular Hollow Section (CHS) Joints under Predominantly Static Loading*, TÜV-Verlag, Köln, Germany

88. Watanabe, E., Sugiura, K., Yamaguchi, T. (1997), Experimental Study on High-Strength Bolted Tube-Flange Joints, *Proceedings*, $5^{th}$ International Colloquium on Stability and Ductility of Steel Structures, Nagoya, Japan, Vol. 1, pp 847-854
89. Wheeler, A.T., Clarke, M.J. and Hancock, G.J. (1995), Tests of Bolted End Moment End Plate Connections in Tubular Members, *Proceedings*, Fourteenth Australasian Conference on the Mechanics of Structures and Materials, Hobart, Tasmania, University of Tasmania, Tasmania, Australia, pp 331-336
90. Wheeler, A.T., Clarke, M.J., Hancock, G.J., and Murray, T.M. (1997), Design Model for Bolted Moment End Plate Connections using Rectangular Hollow Sections, *Research Report R745,* Department of Civil Engineering, The University of Sydney, Sydney, Australia
91. Wheeler, A.T., Clarke, M.J., Hancock, G.J., and Murray, T.M. (1998), Design Model for Bolted Moment End Plate Connections Joining Rectangular Hollow Sections, *Journal of Structural Engineering*, ASCE, 124(2), pp 164-173.
92. Wheeler, A.T., Clarke, M.J. and Hancock, G.J (2000), FE Modeling of Four Bolt Tubular Moment End-Plate Connections, *Journal of Structural Engineering*, ASCE, 126(7), pp 816-822
93. Wheeler, A.T., Clarke, M.J. and Hancock, G.J. (2003), Design Model for Bolted Moment End Plate Connections joining Rectangular Hollow Sections using Eight Bolts, *Research Report R827,* Department of Civil Engineering, The University of Sydney, Sydney, Australia
94. Wilkinson, T.J. (1999), The Plastic Behaviour of Cold-Formed Rectangular Hollow Sections, *PhD Thesis*, The University of Sydney, Sydney, Australia
95. Wilkinson, T.J. and Hancock, G.J. (1999), Comparison of Analyses with Tests of Cold-Formed RHS Portal Frames, In: *Mechanics of Structures and Materials*, Bradford, M.A., Bridge, R.Q. and Foster, S.J. (eds), Balkema: Amsterdam, pp 245-250
96. Wilkinson, T.J. and Hancock, G.J. (2000), Tests to Determine Plastic Behaviour of Knee Joints in Cold-Formed RHS, *Journal of Structural Engineering*, ASCE, 126(3), pp 297-305
97. Wilkinson, T.J., Petrovski, T., Bechara, E. and Rubal, M. (2002), Experimental Investigation of Slot Lengths In RHS Bracing Members, In: *Advances in Steel Structures*, Chan, S.L. Teng, J.G and Chung, K.F. (eds), Elsevier Science Ltd: Oxford, pp 205-212
98. Willibald, S., Packer, J.A., Martinez Saucedo, G. and Puthli, R.S. (2004), Shear Lag in Slotted Gusset Plate Connections to Tubes, *Proceedings*, Fifth ECCS/AISC Workshop on Steel Connections, Session 8, Paper No. 3, Amsterdam, The Netherlands
99. Willibald, S., Puthli, R.S. & Packer, J.A. (2003), Investigation on Hidden RHS-Joint Connections under Tensile Loading, In: *Tubular Structures X*, Jaurrieta, M.A., Alonso, A. and Chica, J.A. (eds), Balkema: Rotterdam, pp 217-225
100. Wright, H.D. (1995), Local Stability of Filled and Encased Steel Sections, *Journal of Structural Engineering,* ASCE, 121(10), pp 1382-1388
101. Zhao, X.L. and Hancock, G.J. (1991), Tests to Determine Plate Slenderness Limits for Cold-Formed Rectangular Hollow Sections of Grade C450, *Steel Construction*, Journal of the Australian Institute of Steel Construction, 25(4), pp 2-16
102. Zhao, X.L. and Puthli, R.S. (1998), Comparison of SCF Formulae and Fatigue Strength for Uniplanar RHS K-Joints with Gap, *IIW Doc. XV-E-98-235*, Monash University, Melbourne, Australia

103. Zhao, X.L., Herion, S., Packer, J.A., Puthli, R.S., Sedlacek, G., Wardenier, J., Weynand, K., van Wingerde, A.M. and Yeomans, N. (2000), *Design Guide for Circular and Rectangular Hollow Section Welded Joints under Fatigue Loading*, TÜV-Verlag, Köln, Germany
104. Zhao, X.L. (2000): Section Capacity of Very High Strength (VHS) Circular Tubes under Compression, *Thin-Walled Structures*, 37 (3), pp 223-240
105. Zhao, X.L. and Grzebieta, R.H. (2002), Strength and Ductility of Concrete Filled Double Skin (SHS inner and SHS outer) Tubes, *Thin-Walled Structures*, 40(2), pp 193-213
106. Zhao, X.L., Grundy, P. and Han, L.H. (2001): Behaviour of Grouted Sleeve Connection at Elevated Temperature, *Proceedings*, In: *Tubular Structures IX*, Puthli, R.S. and Herion, S. (eds), Balkema: Lisse, The Netherlands, pp 453-460
107. Zhao, X.L., Grzebieta, R.H. and Elchalakani, M. (2002a), Tests of Concrete-Filled Double Skin CHS Composite Stub Columns, *Steel & Composite – An International Journal*, 2(2), pp 129-146
108. Zhao, X.L., Grzebieta, R.H., Ukur, A. and Elchalakani, M. (2002b), Tests on Concrete-Filled Double Skin (SHS outer and CHS inner) Composite Stub Columns, In: *Advances in Steel Structures*, Chan, S.L. Teng, J.G and Chung, K.F. (eds), Elsevier Science Ltd: Oxford, pp 567-574
109. Zhao, X.L., Binh, D.V., Al-Mahaidi, R. and Tao, Z. (2004), Stub Column Tests of Fabricated Square and Triangular Sections Utilizing Very High Strength Steel Tubes, *Journal of Constructional Steel Research.*, 60(11), pp 1637-1661

# Subject Index